Geological Treasures of Red Rock Canyon National Conservation Area and Spring Mountain Ranch State Park
With a Chapter on Archaeology

by
Marvin (Nick) Saines, Ph.D.
Illustrations by Tony Ortiz

© 2022 by Marvin Saines. All rights reserved.

Note to Reader: The author welcomes feedback. Please let me know any errors, suggestions, corrections, or new findings.

Marvin (Nick) Saines greatunc@aol.com

Dedication

This book is dedicated to

John Peck
and
Harold Larson

Two remarkable men who have led exemplary lives, and who have made significant contributions to the sciences of geology and archaeology in Southern Nevada.

TABLE OF CONTENTS

INTRODUCTION..1

ACKNOWLEDGEMENTS...9

SUMMARY OF THE GEOLOGY OF RED ROCK CANYON NATIONAL CONSERVATION AREA (NCA) AND SPRING MOUNTAIN RANCH STATE PARK..............................12

PALEOZOIC ERA...23
    Waterfall Canyon – Truly a Treasure...............23
    Turtlehead Peak..26
    Fossil Ridge – Fossils in the Permian Kaibab
        Formation..28
    The Permian-Triassic Contact........................32
        "The Great Dying"..............................32

MESOZOIC ERA – TRIASSIC PERIOD......................34
    Timpoweap Conglomerate and Breccia............34
        Channels of Timpoweap Conglomerate
            Incised into Kaibab Formation...............37
    Virgin Limestone – The Last Marine Formation in
        Southern Nevada..41
    Chinle Formation – Three Intriguing Geological
    Units, Including a Feature of Unknown Origin..44
        The Conglomerate Unit of the
            Shinarump Conglomerate...................44
            The Grotto..............................48
        Petrified Wood in the Sandstone Unit of
            the Shinarump Conglomerate.............50
        Feature of Unknown Origin in the
            Sandstone Unit of the Shinarump

Conglomerate................................52
Petrified Forest Member......................54
   Metoposaur Vertebrae Found in
     Spring Mountain Ranch............56

MESOZOIC ERA – JURASSIC PERIOD......................58
  Lower Jurassic Kayenta-Moenave Formation –
   What Happened to the Fossils?......................58
    Siliceous Layer in Kayenta Formation –
     Derived from Volcanoes in
     California?..62
    Carbonate Cap in Kayenta on SMYC
     Trail..66
  Volcanic Deposits in Spring Mountain
   Ranch?..68
  Aztec Sandstone...69
    The Majestic Wilson Cliffs....................69
    The Great Jurassic Sand Sea.................72
    Cross-Bedding.................................73
    Origin of the Red Color of the Rocks......75
    Other Rock Colors............................76
    Iron Oxide Concretions –"The Rocks
     Have Measles"................................78
  Discovery of Dinosaur Footprints in Red Rock
   Canyon..80
    Dinosaur Tracksites #1 and #2............81
     Hikers Discover the First Dinosaur
      Tracks (Site #1)......................81
     A Geologist Arrives at Red Rock
      Canyon................................81
     Confirmation of the Dinosaur
      Tracks by the Geologist.............83
    Discovery and Origin of the Iconic

  Dinosaur Track (Site # 2).....................85
  Identification of the Tracks by a Paleon-
   tologist..87
  Verification and Mapping by BLM
   Paleontologist...................................89
  Dinosaur Tracksite #3 - Near Pine
   Creek Canyon..................................89
  Dinosaur Tracksite #4 - The Calico Hills
   Site...92
  Dinosaur Tracksite #5 - Pine Creek
   Canyon..92
  Dinosaur Tracksite #6 - The Willow
   Springs Site......................................94
  "Dinosaurs Don't Give Up Their Secrets
   Easily"..95

MESOZOIC ERA – CRETACEOUS PERIOD..................98
  Conglomerate of Brownstone Basin...............98

CENOZOIC ERA..101
  Ancient Landslides.......................................101
   Blue Diamond Landslide.....................102
   Turtlehead Landslide.........................104
  Potato Knoll – A Slump Block.....................106
  Recent Debris Flows..................................108
  Natural Dam in Juniper Canyon....................110
  Skull Rock Landslide North of Pine
   Creek..113
  Precariously Balanced Rocks........,............115
  Spring Deposits in Calico Basin...................117
  Desert Varnish – "The Rocks Are Alive".........119

RED ROCK CANYON FAULTS..................................121
    The Keystone Thrust Fault – Red Rock Canyon's
        Famous Fault...121
    La Madre Fault and a Reversal of
        Topography...125
    The Red Spring Thrust Fault......................127
        The Fenster – A Window Through the
            Red Springs Thrust Sheet................128
    The Cottonwood Fault – See it on the Road to
        Pahrump (Route 160)..............................130

ARCHAEOLOGY – NATIVE AMERICAN ARTIFACTS.....132
    Petroglyphs – Carved in the Rock................132
        Petroglyphs Related to the Old Spanish
            Trail...133
    Pictographs – Paintings on the Rocks...........138
        Brownstone Canyon.......................138
        Willow Springs...............................140
    Agave Roasting Pits..................................142
    Rock Shelters..144
    Please Don't Forget................................144

REFERENCES..146

ABOUT THE AUTHOR..150

INDEX...151

INTRODUCTION

This book was written for visitors to Red Rock Canyon National Conservation Area (RRCNCA) and Spring Mountain Ranch State Park, located less than 20 miles west of Las Vegas, Nevada. Geographically and geologically, Spring Mountain Ranch, and Bonnie Springs development (The Reserve at Red Rock Canyon), are part of Red Rock Canyon. The RRCNCA extends from south of Route 160 north to the vicinity of Kyle Canyon and covers almost 200,000 acres. Spring Mountain Ranch covers 528 acres and is located north of Route 160 and south of the Scenic Drive.

This book is about the geological features in the southern portion of Red Rock Canyon from the vicinity of Route 160 to Brownstone Canyon (Figure 1). This is the area frequented by visitors, hikers, bikers, and horseback riders. When the general geology of Red Rock Canyon is discussed, it includes Spring Mountain Ranch, which has the same rock strata. Site specific places in Spring Mountain Ranch are noted, all other locations are in the National Conservation Area (NCA).

New discoveries discussed in this book include Triassic channels cut into the Permian limestone, dinosaur footprints in the Jurassic sandstone, vertebrate fossils in the Triassic Chinle Formation, the siliceous layer and the carbonate cap in the upper Kayenta Formation, the relationship of Precariously Balanced Rocks to seismic activity in Red Rock Canyon, and the natural dam in Juniper Canyon.

A summary of the geology of Red Rock Canyon and

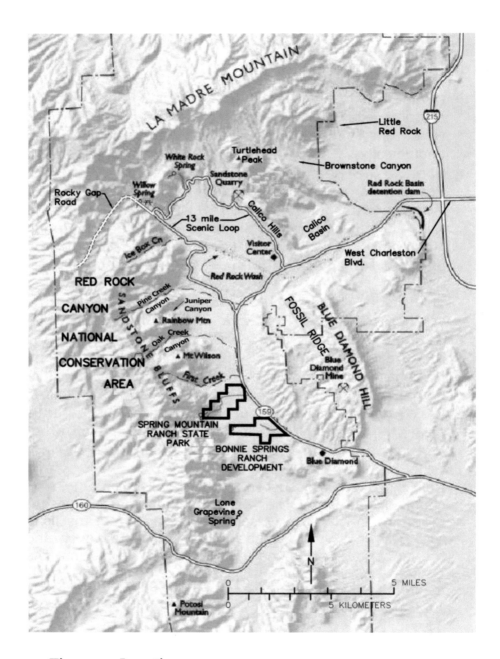

Figure 1 - Location map.

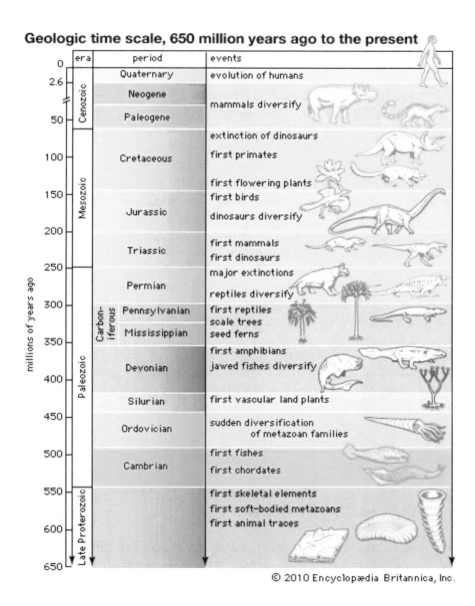

Figure 2 - Geologic time scale. Source: Encyclopedia Brittanica.

Spring Mountain Ranch is presented. Photographs of features that are not described in detail in the following chapters are included in the summary.

The features related to the geological formations are described in geological chronological order (see Figure 2). That is, they start with the features related to the Paleozoic Era of geological time beginning with Waterfall Canyon, in which the rocks are Cambrian in age; then the Mesozoic Era, including dinosaur footprints; and then the Cenozoic Era, including landslides and debris flows. Following the descriptions of features related to the geologic formations, the major faults are discussed, including the world-class Keystone Thrust Fault. Native American archaeological features, including petroglyphs and pictographs, are discussed in the final chapter.

Figure 3 is a generalized geologic map of the area, and Figure 4 is a stratigraphic table, showing the geological formations and ages. Figure 5 is a generalized geological cross section through Spring Mountain Ranch, which shows most of the formations discussed in this book and the basic structural geology of Red Rock Canyon, including the National Conservation Area and Spring Mountain Ranch. Figure 6 is a view of Red Rock Canyon from the ridge west of Blue Diamond with the formations and mountains labeled.

The locations of the dinosaur tracks and certain other sensitive sites are not provided. The Bureau of Land Management (BLM), which manages the RRCNCA, has not given permission to reveal these locations in order to prevent vandalism.

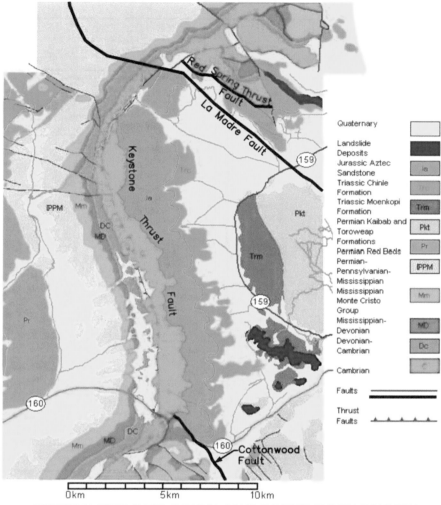

GENERALIZED GEOLOGICAL MAP OF RED ROCK CANYON
MODIFIED FROM DUTCH (2005)
NOTE: L. JURASSIC KAYENTA—MOENAVE FM. MAPPED WITH CHINLE FM.

Figure 3 - Generalized geological map of Red Rock Canyon. This is modified from the map on Professor Steven Dutch's website. The map on the website is simplified from Bohannon and Morris (1983).

Stratigraphic Table for Red Rock Canyon NCA and Spring Mountain Ranch State Park Modified from Tingley and Others (2001) and Axen (1985)			
Geologic Era	Age (Ma)	Geologic Period	Formation Name and Description
Cenozoic	2.6	Quaternary	**Alluvial and debris flow deposits** - Sand and gravel, boulders; caliche.
		Neogene	**Landslide breccia** - Paleozoic limestones and some incorporated
	66	Paleogene	Jurassic sandstone; caps ridges near Blue Diamond and Calico Basin
			Unconformity
Mesozoic		Cretaceous	**Conglomerate of Brownstone Basin** - Pebbles and cobbles of Precambrian, Paleozoic, and Mesozoic rocks, including quartzite
	145		*Unconformity, Lower Cretaceous rocks missing*
		Jurassic	**Aztec Sandstone** - Red and buff cross-bedded sandstone of aeolian (sand dune) origin; makes the massive Wilson Cliffs escarpment
	201		**Kayenta-Moenave Formation** - Brick red siltstone, sandstone, and shale. Moenave, below Kayenta, mostly covered by talus
		Triassic	**Chinle Formation** - **Petrified Forest Member** - Variegated and purple beds of siltstone, mudstone, and clayey shale, some white layers of volcanic ash (?) **Shinarump Conglomerate Member** - Brown ridge-maker; comprised of a red jasper conglomerate unit overlain by a yellow sandstone unit containing petrified wood **Moenkopi Formation** - **Upper Red Member** - Red shale. Weak rock; underlies much of the valley west of Rt. 159 and south of Oak Creek Canyon; mostly covered by alluvium **Virgin Limestone Member** - Gray limestone; ridge-former, few fossils **Lower Red Member** - Gypsiferous red and brown sandstone and siltstone; makes a valley west of Fossil Ridge **Timpoweap Conglomerate Member** - Limestone pebble conglomerate in
	252		channels cut into Permian limestone; basal breccia
			Unconformity
Paleozoic		Permian	**Kaibab and Toroweap Formations** - Gray limestone cliff-formers at top of Blue Diamond Hill; gypsum mined at south end; Kaibab has black chert **Permian Red Beds** - Red siltstone and sandstone underlain by yellow
	299		sandstone, exposed at the north end of Blue Diamond Hill
	323	Pennsylvanian	**Bird Spring Formation** - Gray limestone and dolostone, with sandstone, siltstone, and conglomerate at base
	359	Mississippian	**Monte Cristo Limestone** - Dark gray limestone and dolostone cliff-former
	419	Devonian	**Sultan Limestone** - Gray and black limestone and dolostone cliff-former
			Unconformity. Silurian rocks missing
		Ordovician	**Mountain Springs Formation** - Silty, thin-bedded dolostones; upper member may be Devonian
	485		**Pogonip Group** - Gray dolostone
		Cambrian	**Nopah Formation** - Mostly massive gray to white dolostone and limestone; at base is **Dunderberg Shale Member** - Contains green shale layers **Bonanza King Formation** - Gray dolostone and limestone; lower part
	541		brecciated in fault zones of Keystone Thrust Fault System

Ma means millions of years ago   Saines (2018)

Figure 4 - Stratigraphic table for Red Rock Canyon NCA and Spring Mountain Ranch State Park .

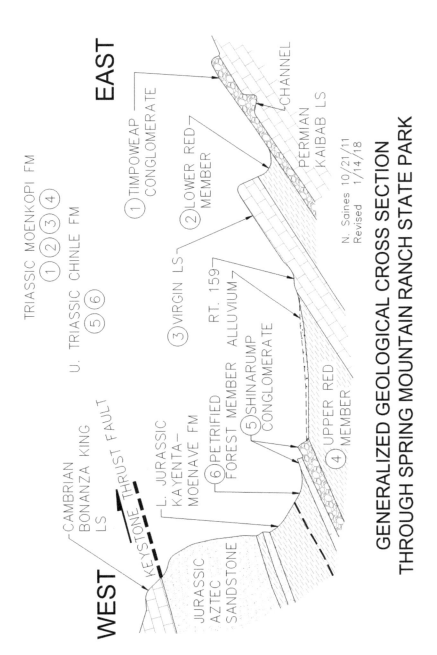

Figure 5 - Generalized cross section through Spring Mountain Ranch. This is the basic stratigraphy (rock layers or formations) and geologic structure underlying Red Rock Canyon, including the National Conservation Area and Spring Mountain Ranch.

Figure 6 - View north from ridge west of Blue Diamond showing geological formations and names of mountains along the escarpment. With Rick Whitaker (left) and John Peck (right).

ACKNOWLEDGEMENTS

When I first began working at Red Rock Canyon in 2011, retired geologist John Peck was my mentor, introducing me to geological features and formations not well known by the casual geologist, such as the Timpoweap Conglomerate. I worked for the Red Rock Canyon Interpretive Association (now the Southern Nevada Conservancy) as a naturalist-geologist from 2011 to 2014.

John and I gradually formed what I call the Geologic Investigative Team – John Peck, Harold Larson, Bill McKinnis, Rick Whitaker, Norma Biggar, Jerry King, and myself. Some of us would go out almost every Thursday during good weather to visit another area and investigate the geology. We are still active, although Norma passed away in 2017. Her book on the geology of Nevada was published the same year (see References under her co-author Frank DeCourten).

Professor Steve Rowland of the Geoscience Department at the University of Nevada – Las Vegas (UNLV) and I have been collaborating on geological projects since the early '90s, when we worked on the Great Unconformity roadside geological park in the northeast part of the Las Vegas Valley. When Gary Smith and Lynn Nicholson showed me the first set of dinosaur prints found in Red Rock Canyon that they discovered in April 2010, I brought Steve to the site, and he identified the tracks as *Grallator* (the track name, not the name of the dinosaur who made the tracks). He went on to work with the BLM on all the dinosaur footprint sites subsequently discovered. These include the one that I found.

Professor Gene Smith of the Geoscience Department at UNLV provided hypotheses for the origin of the siliceous zones found in the Kayenta Formation. Gene also determined through X-ray diffraction that the star-like crystals found in the carbonate cap on the SMYC (Spring Mountain Youth Camp) trail are dolomite. Professor Gerry Bryant of Dixie State University in St. George, Utah, also had input on this issue.

Professor Marjorie Chan of the University of Utah in Salt Lake City made several trips to Red Rock Canyon and provided information on the rock colors in the sandstone, her area of specialization.

Ken Cochran, whom I worked with in Oman in the 1980s, generously provided time and guidance in helping me prepare the first draft on Blurb.com.

I would also like to thank all the hikers on my hikes at Red Rock who had to listen to my explanations and lectures and who pointed out features to look at and explore. Jen Vincent was my assistant leader for over a year. When Jen was promoted and couldn't co-lead hikes any more, Tom Lisby became my assistant leader for over two years. He is now a naturalist and volunteer hike leader for the BLM, and was president of Friends of Red Rock Canyon.

Our regular hikers included Jan Barry, Anita Bouse, Pat and Ralph Cadwallader, Liz Carmer, Pam Carpenter,

Phil Christman, Beth Dixon, Kris Gill, Mike Gottesman, Edwina LaBreque, Curt Pantle, Gareth Pearson, Trish Porter, Jaime Rexing, Duane Siefertson, Pete Stephenson, Clint Wharton, and Cheryl Zentz, now Thomas. Cheryl called us the "Turtleheads," after Turtlehead Peak.

Geologists John Peck, Gary Beckman, and Steve Rowland did the critical technical review and proof-reading of the early drafts. I would like to thank the following for their helpful reviews of the revised version of the book: Nick Christenson, Tom Lisby, David Low, Douglas Sims, and Sali Underwood.

Tony Ortiz did a great job with the maps and cross sections. Aimee and Justin McAffee (Most Media) were able to Photoshop out the geological equipment and dates in several of the photos, thereby saving me the time it would take to go back to re-photograph the rock outcrops. Most Media also produced the front cover of this book.

Most of the photographs in the book are the author's. Thanks go to Liz Carmer, William Cole, Jeff Cuneo, Betty Gallifent, Sendi Kalcic, Harold Larson, Tom Lisby, David Morrow, Gareth Pearson, Lynn Nicholson, Steve Rowland, and Clint Wharton who also contributed photos, as credited in the captions.

SUMMARY OF THE THE GEOLOGY OF RED ROCK CANYON NATIONAL CONSERVATION AREA (NCA) AND SPRING MOUNTAIN RANCH STATE PARK

Note: Photographs of selected features that are not described in detail in the following chapters are included in this Summary.

More than 500 million years ago, the land that would become Red Rock Canyon, including Spring Mountain Ranch State Park, was covered by a shallow sea. Mostly limestone (and dolostone or dolomite) accumulated in this ocean basin for over 250 million years during the Paleozoic Era. The limestones found in Red Rock Canyon contain the fossils of sea life that flourished during that time. The Paleozoic limestones at Red Rock Canyon range in age from Cambrian to Permian. The early- and mid-Paleozoic limestones are exposed on La Madre Mountain on the northwest side of the NCA (Figure 7), and on Turtlehead Peak.

The Cambrian Bonanza King Formation overlies the Jurassic sandstone above the Keystone Thrust Fault. The formation is accessible in Waterfall Canyon on Rocky Gap Road on the west side of the NCA, and at the end of the Keystone Thrust Trail. Early and mid-Paleozoic formations outcrop on Turtlehead Peak, which dominates the scenery north of the Sandstone Quarry area of the Scenic Drive.

The Permian formations outcrop on the east side of

Figure 7 - West end of Calico Hills along Scenic Drive with La Madre Mountain in the background.

Figure 8 - North end of Blue Diamond Hill showing Permian formations. Permian Redbeds correlate with the Hermit Formation. The yellow sandstone unit is equivalent to the Lower Permian Queantoweap Sandstone.

the NCA. At the north end of the Blue Diamond Hill three Permian formations are exposed: the Permian Redbeds, the Toroweap Formation and the Kaibab Formation (Figure 8). The Permian Redbeds underlie the Toroweap Formation. These red shales and sandstones are equivalent to the Hermit Formation of the Colorado Plateau and Grand Canyon. At the base of the redbeds is a yellow sandstone which has been mapped with the Permian Redbeds formation. This sandstone correlates with the Lower Permian Queantoweap Sandstone of western Arizona.

The Toroweap and Kaibab formations are mapped together as the Toroweap-Kaibab Formation by some researchers. The Kaibab Formation, which overlies the Toroweap, is mined for gypsum on the south end of Blue Diamond Hill near the town of Blue Diamond. Permian fossils are found in the Kaibab Formation along the trail on Fossil Ridge near the north end of Blue Diamond Hill.

At the end of the Permian Period (the last period of the Paleozoic Era) there was a mass extinction known as "The Great Dying," in which about 70% of the land vertebrates and 95% of marine life perished (Sahney and Benton, 2008).

Following the Paleozoic Era, about 250 million years ago, the Mesozoic Era began. This was the Age of Reptiles, the Age of Dinosaurs. During the Triassic Period (the first period of the Mesozoic Era), what is now southern Nevada emerged from the sea, and millions of years of mostly continental deposition began. Continental sedimentary rocks are lithified non-marine sediments, deposited by rivers, in lakes, and by the wind. (Lithification is the process of changing unconsolidated sediment into sedimentary rock.)

The oldest Triassic rocks, the Timpoweap Conglomerate Member of the Moenkopi Formation (known as the Rock Canyon Conglomerate in Utah), represents alluvial (river) channels cut into the exposed sea bed.

Above the Timpoweap, the Lower Red Member of the Moenkopi Formation contains red sandstone and shale and gypsum deposits, suggesting an interfingering of non-marine and marine sediments. It is a relatively weak rock and forms a valley between the Timpoweap Conglomerate and the Virgin Limestone (Figure 9).

The sea returned to the region, and the Virgin Limestone Member of the Moenkopi Formation was deposited. As this was after the Great Dying, the formation is relatively devoid of fossils compared to the Permian Kaibab Formation. This was the last advance of the sea into what is now Southern Nevada. For the next 225 million years or so, until present time, continental deposition, uplift, and then erosion prevailed.

The Upper Red Member of the Moenkopi, which overlies the Virgin Limestone, is a weak shaly rock that underlies much of the valley east of the great Wilson Cliffs escarpment south of Oak Creek Canyon and west of Route 159 (Figure 10). It is mostly covered by alluvium and debris flow deposits. These deposits are sand and gravel and boulders deposited during flood events in washes that drained the escarpment to the west. The Upper Red Member of the Moenkopi Formation represents non-marine deposition in rivers and lakes.

Above the Upper Red Member of the Moenkopi Formation is the Shinarump Conglomerate Member of the Chinle Formation, a ridge-former that can be traced from Route 160 north to Oak Creek Canyon. It also outcrops west of the White Rock Hills and in the Calico Basin. The Shinarump represents an apron of coarse alluvium from a mountain range to the southeast that was uplifted and eroded away during Triassic time. The sandstone unit of the Shinarump, above the conglomerate, contains petrified wood and logs, indicating humid conditions during upper Shinarump time.

Overlying the Shinarump Conglomerate is the Petrified Forest Member of the Chinle Formation, which is best exposed in Spring Mountain Ranch. This shaly formation represents lake and swamp deposits. *Metoposaur* vertebrae fossils were found in the outcrop at the Ranch. A *Metoposaur* is an extinct crocodile-like reptile that inhabited the swampy lakes.

Above the Chinle Formation is the Kayenta-Moenave Formation, predominantly red siltstones, sandstones, and shales that were formed in freshwater lakes and rivers. In Red Rock Canyon the Kayenta and Moenave formations have been mapped together, and the lower portion, the Moenave equivalent, is mostly covered by talus. While rich in dinosaur fossils in Utah, no fossils have been found in the Kayenta-Moenave Formation in Red Rock Canyon. An unusual siliceous layer, possibly of volcanic origin, is found in places in the upper part of the Kayenta Formation. A carbonate cap in the upper Kayenta is exposed on the SMYC trail between Ice Box Canyon and Lost Creek Canyon.

Figure 9 - Looking west from Fossil Ridge. Lower Red Member of Moenkopi Formation forms a valley east of the Virgin Limestone.

Figure 10 - Geologist John Peck at outcrop of Upper Red Member of Moenkopi Formation north of Oak Creek. Weak red shale is overlain by alluvium and debris flow boulders.

The great sandstone cliffs at Red Rock Canyon, thousands of feet high, are made up of the Aztec Sandstone. The escarpment is known as The Wilson Cliffs. The Aztec Sandstone, about 180-190 million years old at its contact with the underlying Kayenta Formation, is comprised of lithified sand dunes that formed in a vast desert that covered a large part of what is now the southwestern United States during Jurassic time. It is known as the Great Jurassic Sand Sea (Blakey and Ranney, 2008).

Massive cross-bedding, typical of aeolian (wind) deposits, is a prominent feature of the Jurassic sand dune deposits. Lithification occurred when the natural cements dissolved in subsurface water percolating through the sediments precipitated in the pore spaces between the grains, changing the sand into sandstone. The cements are iron oxide, silica, and calcium carbonate. Subsequently, over millions of years, the rocks were uplifted thousands of feet to their present elevation, tilted gently to the west, and exposed to weathering and erosion.

The red color of many exposures of the Aztec Sandstone is due to the presence of hematite, a form of iron oxide. The distribution of the iron oxide in the rocks is not fully understood by geologists. Other rock colors are due mainly to other iron minerals besides hematite. Exposure to the elements caused some of the iron minerals to oxidize or "rust," or become hydrated, resulting in red, orange, and brown-colored rocks. Areas where the rock is buff in color may be places where the iron has been leached out by subsurface water, or where the iron oxide was never deposited.

Red or brown spots and balls are iron concretions, where subsurface water has precipitated iron oxide around a nucleus in the sandstone. These concretions are more resistant to erosion than the surrounding sandstone, and weather out in places into little balls known as Indian or Moqui Marbles.

In the fall of 2011 the BLM announced the discovery of dinosaur tracks at Red Rock Canyon. The tracks of small, tridactal (three-toed), bipedal (two-footed), therapod (meat-eating) dinosaurs have been found in the Aztec Sandstone. The tracks have been identified as *Grallator*. *Grallator* is the name of the track, not the track-maker. The dinosaur that made the track is probably *Coelophysis* or *Megapnosaurus* - a dinosaur the size of a large dog. Small animal tracks made by proto-mammals, early mammals, and arachnids were also found. Six dinosaur footprint sites found at Red Rock Canyon are described in this book.

In the NCA, a thin layer of the Cretaceous Conglomerate of Brownstone Basin is found in Brownstone Canyon, and in the fenster or window at the end of Keystone Thrust Trail (see below). It is made up of Precambrian, Paleozoic, and Mesozoic pebbles and cobbles, and was deposited on an erosion surface cut into the top of the Aztec Sandstone. It correlates with the Willow Tank Formation, a conglomerate in the Valley of Fire State Park, located about 50 miles northeast of Las Vegas.

There are four major faults in Red Rock Canyon. As you look at the escarpment to the west you see older gray Paleozoic limestone resting on top of the younger Jurassic sandstone. The occurrence of older rock on top of younger rock is a result of the Keystone Thrust Fault. During the Mesozoic Era the oceanic Farallon Plate began subducting (moving down into the mantle) beneath the western edge of the North American Plate. As a result of this convergent boundary subduction, the Sierra Nevada granite batholith was intruded. Compressional forces in the Earth's crust were dominant, and the older rock was thrust up over the younger rock, over millions of years. The compressional thrust faulting at the end of the Mesozoic Era, known as the Sevier Orogeny, can be traced all the way up into Canada. However, one of the best exposures of the thrust faulting in the entire thrust belt is here in Red Rock Canyon.

Another important fault and major feature of the Red Rock Canyon landscape is the La Madre Fault. The near-vertical La Madre Fault trends northwest-southeast across the northern part of the NCA, and can be traced for more than 30 miles across the Spring Mountains. It is responsible for the abrupt outcrop of the Aztec Sandstone at the Calico Hills along the Scenic Drive.

The Red Spring Thrust Fault is part of the Keystone Thrust Fault System. The fault is beautifully exposed in the northwest part of the Calico Basin where the limestone/sandstone fault contact is clearly seen. At the end of the Keystone Thrust Trail the fenster is a window eroded through the overthrust limestone into the underlying younger sandstone.

Another important vertical fault is the Cottonwood Fault along Route 160. The Mesozoic sandstone escarpment ends at the fault with Paleozoic limestone of the Keystone overthrust plate down-dropped south of Route 160. So you have Mesozoic sandstone to the north of the fault and Paleozoic limestones south of the fault.

Another major feature of the geology of Red Rock Canyon is the occurrence of two enormous Cenozoic Paleogene or Neogene landslide deposits. One is located near Blue Diamond, the Blue Diamond Landslide, and the other is located north of the Calico Basin, and is known as the Turtlehead Landslide. After the Keystone Thrust occurred, the limestones perched at the east edge of the top of the sandstone escarpment were unstable and came down to the east into the Valley as huge landslides, incorporating blocks of the underlying sandstone with it in places.

More recent Cenozoic (Quaternary) features include a huge slump block, debris flows, landslides, precariously balanced rocks, and desert varnish. Potato Knoll is a slump block that slid out from the escarpment on the weak Petrified Forest Member of the Chinle Formation. Recent debris flows and landslides coming out of the canyons and off the cliffs are found in places along the escarpment. The mouth of Juniper Canyon is partially blocked by landslide deposits.

The occurrence of numerous "Precariously Balanced Rocks" or PBRs is interpreted to mean that Red Rock Canyon is not in a seismically active zone. PBRs are rocks that look as though they would be dislodged in even a moderate earthquake.

Desert varnish is a black patina on the sandstone rock that forms slowly over hundreds or thousands of years. It is formed by manganese and iron-fixing bacteria generally on smooth vertical cliff faces and on bedding planes in the sandstone.

Recent deposits include wash terraces and wash channels. These features are not covered in this book.

Archaeological features left by Native Americans in Red Rock Canyon include petroglyphs, pictographs, agave roasting pits, and rock shelters. Petroglyphs were etched into the desert varnish. Pictographs were painted on the rock using natural pigments. Agave roasting pits were used by generations of Native Americans for hundreds if not thousands of years to cook food. And rock shelters are natural cave-like openings in the rock in which the Native Americans lived for periods of time.

PALEOZOIC ERA

Waterfall Canyon - Truly a Treasure

One of the most beautiful and unique places in Red Rock Canyon is a little-known gem called Waterfall Canyon (Figure 11). It is located about one mile up Rocky Gap Road from Willow Springs, where water flows across the road almost all year long. At the mouth of the canyon, along the road, is an exposure of the Keystone Thrust Fault, including fault breccia (crushed and broken rock). There is also colluvium down slope of the outcrop. Colluvium is made up of unconsolidated deposits derived from erosion of the rock, carried down slope by water and gravity.

The trail begins where the Rainbow Mountain Wilderness boundary begins. Here the oldest rock formation in the National Conservation Area is found – the Cambrian Bonanza King Formation. This carbonate (limestone and dolostone) formation is over 500,000,000 years old at its base. The more resistant layers of rock in the formation form the waterfalls.

Waterfall Canyon has a series of waterfalls one after the other as you hike up the trail and climb the rocky waterfalls. They start about three feet high, then five feet high, then ten feet high, until, less than a mile up the canyon, you come to a series of waterfalls about 30 feet high. The trail ends at an unclimbable waterfall, but by climbing up the ridge on the north side of the canyon one can see that the waterfalls continue to the west farther up the canyon.

Figure 11 - Two views of thirty-foot high waterfall in Waterfall Canyon. The water is derived from groundwater seepage upstream plus rainfall runoff and seasonal snowmelt.

Figure 12 - Feature in Cambrian Bonanza King Formation intepreted to be a stromatolite. See pen for scale.

The gray limestone and the water create a strange glow in the canyon. The light seems different and beautiful.

What I am interpreting as a fossil *stromatolite* has been found in the rock (Figure 12), less than ½ mile in from the trailhead. These algal mats are some of the earliest life forms found fossilized in rock – and they are still forming today in shallow, warm, marine environments such as in Bermuda and the Bahamas!

Just below the last waterfall, at the end of the trail, springs and seeps emerge from the valley walls that provide some of the water in the channel. There is a reason for everything in nature. Someday a geologist or geology student will discover why the seeps are located there and will address the larger question which is why there is so much water flow in Waterfall Canyon and how much is derived from groundwater and how much is surface water runoff and snowmelt. When it hasn't rained or snowed for months it is obvious that the water flowing in the canyon is derived from groundwater seepage.

Turtlehead Peak

Turtlehead Peak has almost cult status among the Red Rock hikers and visitors because of its majestic appearance and its reputation as a challenging hike. It is one of the most difficult of the 26 hikes listed in the Red Rock Canyon Visitor Guide (2,000 ft. gain in elevation and five miles round trip). Viewed from the east, the pointy peak resembles a turtle's head (Figure 13). (A biologist might point out that it should have been named Tortoisehead Peak, since we have the desert tortoise here in the Mojave Desert. Desert-dwelling turtles are referred to as tortoises.)

The mountain is underlain by early Paleozoic carbonates - limestone and dolostone (or dolomite). These include Cambrian and Ordovician formations capped by the resistant Devonian Sultan Limestone. The Silurian is missing here (Axen, 1985). See the stratigraphic table (Figure 4).

These Paleozoic carbonate formations were thrust over the Jurasssic Aztec Sandstone as part of the Keystone Thrust System (see below), locally mapped as the Red Spring thrust fault. Also see Figures 87 and 88.

Figure 13 - Turtlehead Peak, looking northwest from saddle on Five Stop Hill. Axen (1985) mapped the fault east of the peak as the "Turtlehead Mountain Fault."

Fossil Ridge – Fossils in the Permian Kaibab Formation

Paleozoic limestones and dolomites are exposed in three places in Red Rock Canyon: in the northwest part of the National Conservation Area, visible from the Visitor Center (Figure 7), in the limestone cap of the Keystone Thrust Fault System, including Turtlehead Peak (Figure 13), and on Blue Diamond Hill (Figure 8). The western part of the northern end of Blue Diamond Hill is known as Fossil Ridge (Figure 1). The rocks exposed here are upper Paleozoic carbonates.

Fossil Ridge has three interesting geological features: Permian fossils, the Triassic/Permian contact, and Triassic channels cut into the Permian limestone. The latter two are discussed in separate sections, below. Fossil Ridge is located about one mile south of the entrance to the fee booths on Route 159. The trail up Fossil Ridge is accessed from the Cowboy Trail Rides parking lot. The "green map," a map of Red Rock trails on a topographic base map (for sale at Elements gift shop in the NCA Visitor Center) identifies Fossil Ridge as the hill northeast of the Cowboy Trails parking lot. While this hill is in the same formation (the Permian Kaibab Formation) and has fossils, Fossil Ridge is actually the long ridge alongside (west of) Echo Canyon (Figure 1).

The Kaibab Formation contains Permian fossils, including brachiopods, crinoids, bryozoans, corals, and sponges. These are animals that lived in the Permian sea more than 250 million years ago (see Figure 14). Except for the silicified circular skeletons of sponges

(Figure 14B) the fossils are not that easy to spot. One needs to slowly walk on the rocks and look carefully. After a while they become recognizable. *No fossil collecting is permitted in a National Conservation Area without a special permit.*

The Kaibab limestone contains black chert nodules (Figure 15). The chert is black in color due to a coating of desert varnish. Broken pieces reveal, however, that it is gray inside. There are also many places where black chert appears to have invaded the limestone. This may have occurred during "diagenesis" (Nielson, 1983). Diagenesis is defined as physical and chemical changes that occur after the sediment is deposited, but before it is lithified (turned to rock). However, the current intepretation of the origin of the chert is that it is secondary, precipitated by circulating silica-rich groundwater after lithification.

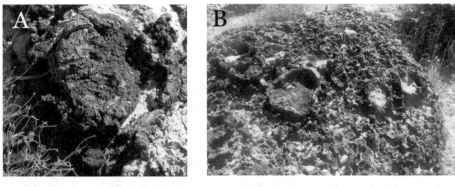

Figure 14 - Permian fossils on Fossil Ridge.
A - Sponge. About 5 inches in diameter.
B - Sponge skeletons, up to 6 inches in diameter.
C - Brachiopods. Up to 1.0 inch in diameter.
D - Crinoid stems and plates. Round plates about 0.25 inch in diameter, stems up to 1.0 inch long.
Photos B and C by Liz Carmer.

Figure 15 - Kaibab Formation showing black chert nodules in six foot high outcrop. The view is north, from just north of the beginning of the Scenic Drive, northeast of the Red Rock Canyon Visitor Center. Aztec Sandstone across the La Madre Fault in the background.

The Permian-Triassic Contact

"The Great Dying"

The greatest mass extinction in geologic history occurred at the end of the Permian Period, about 250 million years ago. About 95% of marine life and 70% of terrestrial life perished (Sahney and Benton, 2008). This is known as "The Great Dying." (The end of Cretaceous extinction that killed the dinosaurs was 66 million years ago, about 184 million years after The Great Dying.)

Scientists do not agree on the cause of the Great Dying, which is still under investigation. There are many hypotheses for its origin. One hypothesis is that volcanic eruptions in Siberia for hundreds of years put so much carbon dioxide into the atmosphere that the Earth warmed up over 5° C. (9° F.). This caused methane hydrate deposits on the oceanic continental shelves to release the methane. Methane killed sea life on the continental shelves. With the addition of methane, another "greenhouse gas" that causes global warming, the Earth warmed up 15° C. (27° F.), which was enough to kill 95% of marine life and 70% of terrestrial life on the planet.

Another hypothesis is that a key factor in the Permo-Triassic extinction was the abundance of sulfur dioxide in the Siberian eruptions. Sulfur dioxide combines with water to form sulfuric acid, which is deadly to life.

The Triassic Virgin Limestone Member of the Moenkopi Formation, which is the next significant marine layer above the Permian Kaibab Formation, is relatively devoid of fossils in Red Rock Canyon compared to the Kaibab, due, in part, to the mass extinctions at the end of the Permian Period (see below).

MESOZOIC ERA - TRIASSIC PERIOD

The Timpoweap Conglomerate and Breccia

The Triassic Timpoweap Conglomerate is a river deposit that formed when the Permian sea bed emerged due to lowering of sea level or rising of the land. The area that is now Southern Nevada was under the sea for over 250 million years. When the sea bed emerged in the early Triassic, rivers began cutting into it, tearing it up. The base of the Timpoweap Conglomerate in Red Rock Canyon is a breccia, with angular hunks of the underlying Kaibab limestone in the matrix of the breccia (Figure 16). The Timpoweap Conglomerate is the lowermost member of the Triassic Moenkopi Formation. (In Utah it is known as the Rock Canyon Conglomerate.)

Several outcrops of the Timpoweap Conglomerate are found on the north end of Blue Diamond Hill, including Fossil Ridge. The outcrop area along the trail on top of Fossil Ridge, where the breccia of Figure 16 is exposed, is said to be a habitat for bobcats, although I personally have never seen one there. The Timpoweap outcrops are remnants of channels cut into the exposed sea bed (see below). Another excellent exposure of the Timpoweap

Figure 16- Breccia at the base of the Triassic Timpoweap Conglomerate containing large angular blocks of underlying Permian Kaibab limestone. This outcrop is near the top of the trail on Fossil Ridge. Blocks are up to three feet in diameter.

Figure 17 - "The Muffins" at the north end of Blue Diamond Hill. The outcrop is made up of Triassic Timpoweap Conglomerate. View looking north. Calico Basin and La Madre Mountain visible in the background. Outcrop is about 15 feet high.

Conglomerate is in "The Muffins" outcrop on the north end of Blue Diamond Hill (Figure 17). An easily accessible (except for the barbed wire) outcrop of the Timpoweap is at the junction of Route 159 and the turnoff to the village of Blue Diamond, on the east side of the road.

Channels of Timpoweap Conglomerate Incised into Kaibab Formation

Clearly visible on Fossil Ridge and to the east on the north end of Blue Diamond Hill are the Triassic conglomerate channels incised into the Permian limestone when the land emerged out of the sea at the end of the Permian Period (Figures 18 and 19). According to Blakey and Ranney's paleogeographic map (Figure 20) the sea was located to the northeast in the early Triassic, so it is expected that the strike or trend of these channels would be generally northeasterly, unless they were tributaries to a northeast flowing channel. The Muffins and the channel depicted in Figure 19 may have been part of the same channel.

Having recognized these Triassic channels cut into the Kaibab, I approached Professor Steve Rowland of the Geoscience Department at UNLV and suggested that mapping these channels would be an interesting Senior thesis for one of his students. A geological investigation of the stratigraphy and sedimentary structures of these deposits, especially cross-bedding, would reveal the direction of flow of the rivers that eroded the channels and deposited the conglomerate. Cross-bedding (Figure 21) forms in the direction of flow. Undergraduate student Oscar Vasquez undertook to map the channels under the supervision of Professor Rowland and myself. Oscar mapped the Timpoweap

Figure 18- Channel of brown Timpoweap Conglomerate cut into gray limestone of Kaibab Formation on Fossil Ridge. Outcrop of the basal breccia of the Timpoweap Conglomerate shown in Figure 16 is near the bottom of the channel. View northwest. Conglomerate channel outcrop is about 30 feet thick.

Figure 19 - View east-northeast from Fossil Ridge trail showing Timpoweap Conglomerate channel cut into Permian Kaibab Formation. Channel is about 35 feet thick.

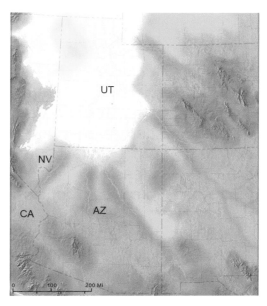

Figure 20 - Paleogeographic map of the southwest during Early Triassic time, about 250 million years ago. The drainage in southern Nevada was northeasterly towards the sea.
Source: Blakey and Ranney (2008).

Figure 21 - Cross bedding in Timpoweap Conglomerate channel on Fossil Ridge. Slope from right to left indicates flow direction, in this case in a northerly direction. Ruler is 10 cm (4 inches) long. Photo by Steve Rowland.

Conglomerate at the north end of Blue Diamond Hill, including Fossil Ridge. Based on the analysis of cross-bedding sets he determined that the direction of flow of the rivers that deposited the conglomerate was northeasterly, in agreement with Blakey and Ranney's paleogeographic map.

Since it is incised into the underlying Kaibab Formation, the Timpoweap Conglomerate, a younger formation, is found at a lower elevation than the top of the Kaibab Formation in places. Not identifying these as channels, previous geologists incorrectly mapped the contact of the two formations as a fault on Fossil Ridge, with the Triassic rock down-dropped below the elevation of the top of the Permian Kaibab Formation.

The Virgin Limestone - The Last Marine Formation in Southern Nevada

In Early Triassic Moenkopi time an arm of the sea advanced into southern Nevada and the Virgin Limestone Member was deposited. This limestone was deposited after the Great Dying, which occurred at the end of the Permian Period, and is relatively devoid of fossils compared to the Kaibab Formation, as mentioned above.

In desert climates limestone is a ridge-former, unlike in the humid eastern U.S. where it is valley-former. The Virgin Limestone ridge is the destination of the Moenkopi Trail, which begins at the Red Rock Canyon Visitor Center and heads west. This moderate trail has no road or cars near it and the visitor is soon out in the desert, heading towards the Virgin Limestone ridge with the magnificent Wilson Cliffs escarpment to the west (Figure 22).

The Virgin Limestone can also be seen as ribs of rock protruding from the ground near the Red Rock Overlook on Route 159, but the best exposure of the formation is east of Route 159 on the west side of Blue Diamond Hill (Figure 9). The formation also outcrops on the west side of Route 159 at Bonnie Springs and to the south.

Figure 22- Google Earth satellite image of northern part of Scenic Drive showing Moenkopi Ridge (Virgin Limestone Member of Moenkopi Formation) and other features. Inset is view west from observation platform in outdoor portion of the Red Rock Canyon Visitor Center.

Figure 23 - Mudcracks and ripple marks in upper Virgin Limestone Member of Moenkopi Formation, just below contact with Upper Red Member. Outcrop is in Oak Creek Wash where it runs parallel to Route 159, north of First Creek trailhead. Hiking poles for scale. Wave length of ripple marks is two to three inches. Photos by William Cole.

The Virgin Limestone Member of the Moenkopi Formation was the last advance of the sea into southern Nevada. Near the top of the formation in the contact zone with the Upper Red Member, hiker William Cole found mudcracks and ripple marks (Figure 23). These features represent the shallow water conditions that prevailed as the environment of deposition changed from marine to non-marine. They are found in places where the Virgin Limestone outcrops east and west of Route 159 between the First Creek parking lot and the parking area along the road at the middle trailhead to Oak Creek. After Virgin Limestone deposition, for the next 225 million years until the present time, continental deposition and then erosion prevailed.

The Chinle Formation – Three Intriguing Geological Units, Including a Feature of Unknown Origin

The three units of the Middle to Upper Triassic Chinle Formation are the conglomerate unit of the Shinarump Conglomerate Member, the sandstone unit of the Shinarump Conglomerate containing petrified wood and logs, and the shaly deposits of the Petrified Forest Member of the Chinle in which vertebrate fossils have been found in Spring Mountain Ranch. The geological feature of unknown origin is in the sandstone unit of the Shinarump. The sandstone unit also contains some conglomerate interbeds.

The Conglomerate Unit of the Shinarump Conglomerate

With the possible exception of the Aztec Sandstone, the conglomerate unit of the Shinarump Conglomerate Member of the Chinle Formation is the hikers' favorite formation. This is because of its beautiful texture and composition, because it forms accessible hikable ridges, and because of its unusual name, which everyone likes to pronounce. (It is named after the Shinarump Cliffs in Utah, the type locality, where it was first described by geologists.)

The formation is comprised of two lithologic (rock) units or *facies* in Red Rock Canyon – the conglomerate and the overlying sandstone. The conglomerate, which appears reddish brown or dark brown from a distance, contains clasts (pebbles) of jasper, a red cryptocrystalline variety of quartz creating a beautiful appearance and texture (Figure 24).

The Shinarump Conglomerate is a ridge-former at the base of Wilson Cliffs escarpment. It can be traced from Route 160 north to Oak Creek Canyon, and outcrops west of the White Rock Hills to the north. Small exposures of the Shinarump Conglomerate are also found in the Calico Basin. The conglomerate represents an apron of coarse alluvium (mostly gravel and sand) from a mountain range to the southeast in the vicinity of New Mexico that was uplifted and eroded away during Triassic time.

The best exposures of the conglomerate are at Spring Mountain Ranch (Figure 25), and the east side of Potato Knoll on the south side of Oak Creek Canyon (Figure 26). At the outcrop west of the White Rock Hills on the White Rock Loop Trail the rocks have been pushed or dragged up almost vertically due to the Keystone Thrust Fault (Figure 27).

Of special interest to geologists are the unusual lithologies (rock types) found in places in the base of the Shinarump above the red shaly Upper Red

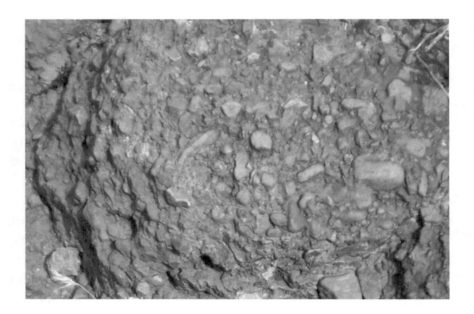

Figure 24 - Closeup of conglomerate facies of Shinarump Conglomerate Member of Chinle Formation showing clasts or pebbles of jasper. Largest clast is about 1.5 inches long.

Figure 25- Shinarump Conglomerate ridge in Spring Mountain Ranch. The outcrop is about 20 feet thick. Figure 24, the closeup of the conglomerate, was photographed at the top of the ridge.

Figure 26 - View west toward Mt. Wilson showing Potato Knoll, and Shinarump Conglomerate ridge that runs along its east side. Conglomerate outcrop is about 15 feet thick.

Figure 27 - Steeply dipping Shinarump Conglomerate on west side of White Rock Loop. The rocks were dragged up and overturned to the west due to over-riding by Lower Paleozoic rocks during the Keystone Thrust Fault events. These carbonates were thrust over the top of the sandstones to the right (east) and have been eroded back to their present position after about 65 million years of erosion. Also see Figures 83 and 84.

Member of the Moenkopi Formation and below the conglomerate ridge. These are thin layers of freshwater limestone and limestone pebble conglomerate. They can be found on the southeast side of Potato Knoll, and in the wash a few hundred feet north of Route 160, just before the road ascends the steep grade to Mountain Springs (Figure 28).

The Grotto

One of the most delightful places to visit in Red Rock Canyon under the right conditions is The Grotto. It is a waterfall and pool in a little basin carved into the Shinarump Conglomerate in First Creek Canyon in the NCA (Figure 29). The "right conditions" means in the winter and early spring during snowmelt and during rainy periods the rest of the year. In extended dry periods the waterfall and pool will be dry. It is located off the First Creek trail about one mile in and a little hard to find on your own. It is best to go with someone who knows where it is. The trail is easy except for descending into the grotto, which is rocky, but not difficult or dangerous.

Figure 28- Geologic Investigative Team examining freshwater limestone and limestone pebble conglomerate below Shinarump Conglomerate (brown unit above) in wash just north of Route 160 as it begins climbing up to Mountain Springs. Closeup of limestone pebble conglomerate to left ; walking stick for scale. John Peck examining outcrop, Bill McKinnis to right, Harold Larson above to right, and Jerry King to left.

Figure 29 - The Grotto, a waterfall and pool in the Shinarump Conglomerate on First Creek. It has water during the spring snowmelt and during rainy periods. Photo by Gareth Pearson.

Petrified Wood in the Sandstone Unit of the Shinarump Conglomerate

Petrified wood and logs are found in the yellow sandstone unit above the conglomerate in the Triassic Shinarump Conglomerate Member of the Chinle Formation. During Chinle time the climate was humid and the region was forested. Conifer trees were knocked down and carried away by rivers and deposited in the sediments, later lithified to rock. Petrified wood forms when mineral-laden groundwater seeping through the deposits replaces the woody tissue cell by cell with silica (quartz), so that the rock retains the appearance of the original wood.

Petrified wood is found in places in the Chinle sandstone unit along the escarpment from Route 160 north to Oak Creek Canyon. The best exposures of petrified wood and logs in the sandstone are on the northeast side of Potato Knoll, where one log is over five feet long (Figure 30).

Before the establishment of the National Conservation Area, petrified wood and logs were quarried out of the sandstone on Potato Knoll. An abandoned wooden sled used to haul logs and a couple of petrified logs removed from the outcrop are located just off the trail on the northeast corner of Potato Knoll. According to Harold Larson (personal communication), petrified wood and logs from Potato Knoll were used in the lobby of a downtown casino in Las Vegas called The Mint. When the Mint was bought by the Binion family and incorporated in the Horsehoe casino the petrified wood and logs in the lobby, 225,000,000 years old, ended up in a landfill.

Figure 30- Petrified log in sandstone unit of Shinarump Conglomerate Member of Chinle Formation. Located in the northeast part of Potato Knoll. Log is 225,000,000 years old. With Harold Larson.

Feature of Unknown Origin in the Sandstone Unit of the Shinarump Conglomerate

The feature shown in Figure 31 is a brown, coarse-grained sandstone lens within the yellow, finer-grained sandstone unit of the Triassic Shinarump Conglomerate Member of the Chinle Formation. It is located on top of the Shinarump Conglomerate ridge in the southeast part of Potato Knoll. The coarser, more competent or brittle, brown sandstone is fractured or cracked, whereas the yellow sandstone is not. This likely occurred when the rocks were uplifted and tilted to the west.

It looks as though the yellow sandstone envelopes the brown sandstone lens. The lens may represent deposition in a higher energy channel within the alluvial (river) deposits. Since there are petrified logs in the yellow sandstone, another hypothesis is that the brown coarse sand was originally deposited in a hollow log. A third hypothesis, suggested by hiker Brenda Pirani, is that the feature is part of a large dinosaur skeleton! She said it resembled a dinosaur skeleton on display in the Royal Tyrrell (Drumheller) Museum in Calgary. It is unlikely that this hypothesis is correct since dinosaurs in the Triassic were relatively small, and the rock does not appear to be related to bone material. Deposition in a higher energy channel is the most likely hypothesis. What do you think?

To determine the feature's origin, more research is

required, perhaps a senior thesis for a college geology student. Note that a collection permit must be obtained from the B.L.M. before any rocks are removed.

Figure 31 - Feature of unknown origin in sandstone unit of Shinarump Conglomerate. The jointed (cracked) brown coarse sandstone appears to be enveloped by the finer-grained yellow sandstone. The brown sandstone outcrop is about ten feet long and up to two feet high. It is located in the southeast part of Potato Knoll.

Petrified Forest Member

The Triassic Petrified Forest Member of the Chinle Formation overlies the sandstone unit of the Shinarump Conglomerate (the one that contains petrified wood and logs). When I first saw the petrified wood and logs in the sandstone overlying the conglomerate in the Shinarump I thought that was the Petrified Forest Member. That is not the case. The Petrified Forest Member, like all geological formations, is named after the "type locality" – the place where it was first described by geologists. This is the Petrified Forest of Arizona. In Petrified Forest National Park the petrified logs and wood are found in the sandstones and conglomerates (river channel deposits) and not in the mudstones (floodplain deposits). No petrified wood has been found in the mudstones of the Petrified Forest Member in Red Rock Canyon.

The Petrified Forest Member of the Chinle Formation is predominantly clayey shale, siltstone, and mudstone, and is best exposed at Spring Mountain Ranch up-slope and north of the Wilson family cemetery, above the Shinarump Conglomerate ridge (Figure 32). Here the rock shows the variegated colors, including gray and purple or lavender, that the formation is known for all over the Southwest. It is also exposed (poorly) in the drainage on the southwest side of Potato Knoll. Some of the thin white layers in the Petrified

Forest Member may be volcanic ash derived from volcanoes in California about 220 million years ago during Late Triassic Time.

Harold Larson and then UNLV paleontology students with permission to explore found fossil vertebrae in the Petrified Forest Member of the Chinle Formation in Spring Mountain Ranch (see below). This formation is also a notoriously weak rock landslide horizon in the region, especially in southern Utah. Potato Knoll just south of Oak Creek, is a slump block that rotated away from the cliff on this unit (see section on Potato Knoll).

Figure 32 - Petrified Forest Member of Chinle Formation as exposed at Spring Mountain Ranch, above the Shinarump Conglomerate ridge.

Note: *Special permission is required to go off trail in Spring Mountain Ranch State Park. Stay on the trail.*

Metoposaur Vertebrae Found in Spring Mountain Ranch

Harold Larson found the first vertebra fossil in Red Rock Canyon at Spring Mountain Ranch State Park in April of 2013. It was tentatively identified by paleontologist Josh Bonde to be a Metoposaur – a Triassic crocodilian that became extinct at the end of the Triassic Period. The Metoposaur, or *Metoposaurus*, was a freshwater amphibian, up to about three meters (ten feet) long, with eyes on the top of its head (Figure 33). It would burrow in the lake bottom and rise to attack animals swimming above it. Dr. Bonde said it would be great to find a skull in order to confirm the identity of the fossil. So far no skull has been found.

The fossil was found in the Petrified Forest Member of the Chinle Formation, above the sandstone unit of the Shinarump Conglomerate Member of the Chinle Formation (Figure 32). Members of the Geologic Investigative Team (the author, John Peck, Harold Larson, Jerry King, Norma Biggar, and Bill McKinnis) were on a geohike at Spring Mountain Ranch, when Harold, a retired engineer, showed his find (Figure 34) to the geologists who doubted it was a fossil. Harold persisted and called in Dr. Joshua Bonde, who was a professor of paleontology at UNLV. Professor Bonde confirmed that it was indeed a vertebra and organized a follow-up field investigation in which more vertebrae bones were found. These are on display in the Las Vegas Natural History Museum.

Remember: Permission is required to go off trail in Spring Mountain Ranch State Park.

Figure 33- Drawing of a Metoposaur, a Triassic crocodilian that lived in freshwater lakes. Source: deviantart.com.
Inset is a photograph of a Metoposaur skull. Source: Wikipedia.

Figure 34 - Metoposaur vertebra found in Petrified Forest Member of Chinle Formation at Spring Mountain Ranch. It is on display at the Las Vegas Natural History Museum.

MESOZOIC ERA - JURASSIC PERIOD

Lower Jurassic Kayenta-Moenave Formation - What Happened to the Fossils?

Above the Chinle Formation is the Kayenta-Moenave Formation, predominantly red siltstones, sandstones, and shales that were formed in freshwater lakes and rivers. In Utah the Kayenta and Moenave are separate formations rich in dinosaur and other fossils. The Dinosaur Discovery Site at Johnson Farm in St. George is in the Moenave Formation. In Red Rock Canyon the formations have been mapped together and the lower portion, the Moenave equivalent, is mostly covered by talus. No fossils have been found yet in the Kayenta-Moenave Formation in Red Rock Canyon. The Utah State Paleontologist Jim Kirkland suggested that perhaps the water was saltier than in Utah. He also said "Keep looking."

The formation is best exposed at Spring Mountain Ranch (Figure 35) and in the wash at the northeast corner of the Calico Basin in the NCA (Figure 36).

When looking at the magnificent escarpment behind Bonnie Springs and Spring Mountain Ranch from Route 159 (Figure 37) the Jurassic Aztec Sandstone/Kayenta Formation contact is readily visible, especially in the morning with the sun illuminating the east-facing cliffs. It is the contact between the thin-bedded red Kayenta shale, siltstone, and sandstone below and the massive buff-colored Aztec Sandstone above. Here it is easy to see a contact between two geological formations.

Although the contact looks sharp and abrupt from Route 159, the geology of the upper Kayenta Formation is complicated. As discussed below there is a siliceous zone or layer and a carbonate cap found in the upper Kayenta Formation. A detailed study of the transition from humid climate rivers and lakes (Kayenta) to arid climate sand dune sedimentation (Aztec Sandstone) in Red Rock Canyon would be a good topic for a master's thesis in geology.

Figure 35 - Jurassic formations at Spring Mountain Ranch, above the Shinarump Conglomerate ridge.

Figure 36 - Red shale, siltstone, and sandstone of the Jurassic Kayenta Formation in the northeast corner of the Calico Basin. Buff-colored Aztec Sandstone above the Kayenta. With John Peck.

Figure 37 - View of contact of Jurassic Aztec Sandstone and underlying Kayenta Formation from Route 159 at entrance to Spring Mountain Ranch.

Another interesting area of research would be why no fossils have been found in the Kayenta in Red Rock Canyon, whereas in Utah it is a prolific fossil-bearing formation. (Also see discussion of carbonate cap in the Kayenta, below).

Siliceous Layer in the Kayenta Formation – Derived from Volcanoes in California?

There are two gnarly (rough-looking) rock units found in the upper Kayenta Formation. One is the siliceous unit found in the Calico Basin, in Pine Creek, and in other places. The second is a carbonate unit that caps an outcrop of red Kayenta sandstone on the SMYC Trail (see below). SMYC stands for Spring Mountain Youth Camp. Boys from the camp provided the labor to build the trail.

Siliceous means containing silica (silicon dioxide) - quartz and other varieties of silica such as chalcedony and opal. On the east end of Kraft Mountain in the Calico Basin, Harold Larson (who discovered the vertebra fossil at Spring Mountain Ranch) noticed a strange-looking gnarly white rock lying on the slope. The Geologic Investigative Team soon found the source layer up-slope near the top of the Kayenta Formation, less than ten feet below the contact with the Aztec Sandstone (Figures 38 and 39). The layer, ranging from zero where it pinches out, to less than three feet thick is more resistant than the red Kayenta shale, siltstone, and sandstone, and stands out in relief. The siliceous layer can be traced all around Kraft Mountain in the Kayenta Formation, but it is in lenses, and not a continuous layer.

The siliceous layer is also found in Pine Creek

Figure 38 - Siliceous layer near top of Jurassic Kayenta Formation in the northeast corner of the Calico Basin. It is less than three feet thick.

Figure 39 - Closeup of siliceous layer near top of Kayenta Formation. Minerals include quartz and chalcedony. Walking stick is less than 1.5 inches in diameter.

Canyon, just above where Kayenta sandstone meets the trail, on the north side of the wash. Here, however, the outcrop is 50 feet or so below the Aztec Sandstone contact. It is also found as "float" (pieces of rock carried by gravity down the slope from an outcrop above) on the south side of Pine Creek along the trail going up to the Arnight Trail, and along the SMYC trail. It is likely that more outcrops of the siliceous layer will be found in Red Rock Canyon in the upper part of the Kayenta Formation.

Professor Eugene Smith of the UNLV Geoscience Department had a thin section made of the unit from the east end of Kraft Mountain and made a preliminary identification of quartz and spherulitic chalcedony in the rock. Spherulitic means elongate crystals radiating from a common center forming a spherule, and chalcedony is a microcrystalline variety of silica. According to Professor Smith, "Its origin is controversial and ranges from recrystallization of biogenic silica (Opal-A) to the devitrification of volcanic glass."

The unit may be derived from silicification that occurred due to subaerial exposure (at the land surface) of the sediments during Kayenta (early Jurassic) time, or the alteration of volcanic ash that was deposited from volcanoes to the west in what is now California during early Jurassic time, about 190 million years ago. The irregular thickness and distribution of the siliceous zone suggest that it may represent an ash fall deposit – absent in some places, accumulating in others. Groundwater seeping through the rhyolitic ash may have precipitated the silicate minerals. In either case the siliceous layer probably represents subaerial exposure of the sediments and subsequent mineralization and alteration.

Carbonate Cap in Kayenta on SMYC Trail

The gnarly carbonate cap is found north of Icebox Canyon about 50 feet below the contact with the overlying Aztec Sandstone (Figure 40). The unit contains stellate (star-like) crystals discovered by hiker Pete Stephenson (Figure 41). Professor Gene Smith had them run on the mass spectrometer at UNLV and they were determined to be dolomite.

The environment of deposition of the Kayenta Formation was rivers and lakes. The carbonates appear to have formed in a Kayenta lake. Dolomites and other carbonates have been reported from modern highly saline lakes in Saskatchewan, Canada (Last and others, 2012). According to Professor Gerry Bryant of Dixie State University in St. George, Utah, there also were shallow lakes and ponds in interdune areas in the Jurassic which were subject to high rates of evaporation, and which produced carbonates. The high salinity could explain why we have not found fossils in the Kayenta in Red Rock Canyon, as suggested by the State Paleontologist of Utah, Jim Kirkland.

Interdune carbonates have also been found in the Jurassic Aztec Sandstone in Valley of Fire State Park, northeast of Las Vegas.

Figure 40 - Carbonate cap in Jurassic Kayenta Formation along the SMYC Trail. Gnarly cap sits on top of typical red rocks of the Kayenta Formation. It is about ten feet thick.

Figure 41 - Star-like dolomite crystals in carbonate cap on the SMYC Trail. Note dolomite stars above the hand and to the left. These may have been deposited in a saline lake during Kayenta (early Jurassic) time.

Volcanic Deposits in Spring Mountain Ranch?

Professor John Marzolf of Southern Illinois University described volcanic deposits in Spring Mountain Ranch in a 1988 Geological Society of America field trip guidebook. He mentions sandstone deposits at the top of the Moenave-Kayenta as being tongues or lenses of the Aztec Sandstone. At the base of the lower of the two tongues are "dark red and purple cobbles and boulders of volcanic rocks, some flow banded.... The single bed of conglomerate, 3 to 4 m thick, is indicative of close proximity to a volcanic source."

The Geologic Investigative Team (see Introduction) was unable to locate this outcrop. Subsequent attempts at locating it with Professors Gene Smith and Steve Rowland of UNLV were also unsuccessful.

Aztec Sandstone

The Majestic Wilson Cliffs

The magnificent sandstone escarpment that rises up to over 3,000 feet above the desert floor in Red Rock Canyon is known as the Wilson Cliffs (Figure 42). It is comprised mainly of Jurassic Aztec Sandstone, 180-190 million years old at its base, which was sand dunes during Jurassic time. They were subsequently lithified into sandstone by cementation of the sand grains by groundwater percolating through the deposits. The Jurassic Period began about 200 million years ago and lasted about 50 million years (Figure 2).

The Wilson Cliffs are named after the Wilson family who had a ranch in the 19th century and early 20th century at what is now Spring Mountain Ranch State Park. Many geologists looking at the steep cliffs assume that the escarpment is due to faulting, but it is actually due to differential erosion. There is no fault along the base of the cliffs (see Figures 3 and 5).

As discussed in sections above, weak shale of the Upper Red Member of the Moenkopi Formation underlies much of the valley east of the cliffs and has mostly been eroded away and covered with alluvium (sand and gravel) carried in from the canyons that cut into the escarpment. Above the

Figure 42 - The Wilson Cliffs escarpment. The cliffs rise 3000 feet above the desert floor to an elevation of over 7000 feet. The Jurassic Aztec Sandstone comprises the main bulk of the cliffs. It is capped by the older Cambrian Bonanza King limestone which was moved up and over the sandstone by the Keystone Thrust Fault.

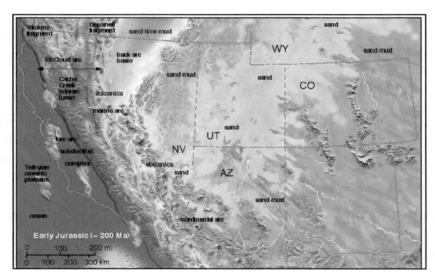

Figure 43 - The great Jurassic Sand Sea, a sand dune desert stretching from present-day California into the Rocky Mountain states that existed from about 200 to 150 million years ago. The Aztec Sandstone is comprised of sand dunes that were deposited at that time and later lithified.
Source: Blakey and Ranney (2008)

Upper Red Member of the Moenkopi Formation at the base of the Wilson Cliffs is the Shinarump Conglomerate Member of the Triassic Chinle Formation, which is more resistant to erosion and makes a ridge. Above the conglomerate and sandstone of the Shinarump is more shale of the Petrified Forest Member of the Chinle, and above that red shale, siltstone, and sandstone of the Lower Jurassic Moenave and Kayenta Formations. Above the red Kayenta shales is the massive Jurassic Aztec Sandstone. The contact of the red Kayenta shales with the overlying buff-colored Aztec Sandstone of the Wilson Cliffs is beautifully exposed looking west from Route 159 near Spring Mountain Ranch (Figure 37).

The Keystone Thrust Fault (see below) resulted in a limestone cap on top of the sandstone, which has eroded back in the last 65 million years, but had served to protect the underlying sandstone from erosion.

The Great Jurassic Sand Sea

Between about 200 and 150 million years ago Southern Nevada was part of an *erg*, a great sand sea, comparable to the Sahara Desert of today (Figure 43). It extended from present-day California into the Rocky Mountain states. It has been called the largest sand dune desert in geological history. The source of the erg is believed to be from winds blowing to the west across great alluvial fans to the east that were derived from westward erosion of the Appalachian Mountains (Blakey and Ranney, 2008).

What is now the Las Vegas Valley and everything around it for hundreds of miles was covered by sand dunes. These sand dunes were inhabited by therapods – meat-eating dinosaurs and other Jurassic animals. Much of the sand was eroded away even before lithification to sandstone. The sandstone was faulted and eroded through time so that today the remnants of the great Jurassic sand sea are found as outcrops of the Aztec Sandstone in the Mescal Range in California, at Red Rock Canyon and Valley of Fire in Nevada, in Zion National Park and in Coral Pink Sand Dunes State Park in Utah, and in other places in the region. In Utah the formation that we call the Aztec Sandstone is known as the Navajo Sandstone. It is called the Nugget Sandstone in the Rocky Mountain states.

Cross-Bedding

A conspicuous feature of the aeolian (wind-blown) Aztec Sandstone is the large-scale cross bedding, which appears as layers of rock dipping at an angle from the horizontal (Figure 44). The cross-bed sets can be huge in size, meters to many meters from top to bottom. The simplest explanation of the different directions of dip is that they are due to shifting wind directions during the Jurassic period, with variable deposition and erosion of the sand.

Cross beds are also found in fluvial (river) deposits, such as the Kayenta sandstone in Pine Creek (Figure 45) and in the Timpoweap Conglomerate on the north end of Blue Diamond Hill (Figure 21). But fluvial cross-bed sets are measured in centimeters instead of meters. Also the angle of the cross bedding in fluvial deposits is generally lower than in aeolian cross bedding.

Figure 44 - Huge cross beds in the Aztec Sandstone. These were formed when the winds shifted direction during Jurassic time when these were sand dunes. Note how the dip or slope of the beds changes direction. The outcrop is over 40 feet high; located near Calico 1 on the Scenic Drive.

Figure 45- Cross bedding in the Kayenta Formation in Pine Creek. These fluvial (river) cross beds are much smaller than the aeolian cross beds shown in Figure 44. The outcrop is about ten feet high.

Origin of the Red Color of the Rocks

The red color of many of the rock units is due to hematite (ferric oxide), an oxidized iron mineral which coats the sand grains and also acts as a cementing agent. (A sandstone is comprised of sand grains cemented together.) Triassic and Jurassic red rocks in Red Rock Canyon, such as the Upper Red Member of the Moenkopi Formation, and the Aztec Sandstone, are non-marine formations. The Triassic Upper Red Member of the Moenkopi Formation was formed in a fluvial (river) or lacustrine (lake) depositional environment, as were the Lower Jurassic Moenave and Kayenta Formations. The Jurassic Aztec Sandstone is comprised of lithified sand dunes formed in a desert environment.

According to Professor Marjorie Chan of the University of Utah in Salt Lake City (personal communication), the Jurassic sand dunes may have originally been red, like the red sand dunes in Namibia and Australia. After deposition, when the deposits were below the water table, the iron may have been leached out in many places by percolating groundwater, changing the deposit to a more buff color.

Other Rock Colors

Buff-colored sandstone was formed by groundwater leaching out the hematite (iron oxide) coating the grains, or the iron was never deposited in that area – the original sand dunes were sand- or buff-colored upon deposition. Rust-colored or brown or yellowish sandstone is usually due to the presence of the hydrated iron oxide minerals – limonite or goethite. A greenish color, like some layers in the Upper Red Member of the Moenkopi Formation, is due to reduced rather than oxidized iron (Figure 46).

In the southeast corner of the fenster (geologic window, see below) rare white-colored sandstone is found. This outcrop is close to a fault, and the lack of color may be due to extensive leaching along the fault zone (Figure 47). The black coating or patina covering the sandstone in many places, known as desert varnish, is discussed below in the Cenozoic section.

Figure 46- Green color due to reduced iron in the Triassic Upper Red Member of the Moenkopi Formation in Oak Creek Canyon. Red color is due to oxidized iron.

Figure 47 - White Aztec Sandstone in the fenster (geological window) at the end of the Keystone Thrust Fault trail. The sandstone is usually buff-colored or red; the white color may be due to extensive leaching by groundwater along a fault.

Iron Oxide Concretions - "The Rocks Have Measles"

Iron oxide concretions are very common in the Aztec Sandstone in Red Rock Canyon. They appear as red or brown bumps, typically about one-inch across and eventually weather out as little balls (Figure 48). These balls are called desert marbles or Moqui marbles, after the Hopi tribe. Native American children would play with these marbles, as city kids used to play with marbles (maybe they still do). Also, the concretions are said to be uitilized in some Native American spiritual activities. The reason that the concretions stand out of the rock is that the iron-impregnated sandstone is more resistant to weathering and erosion than the surrounding sandstone.

Groundwater seeping through the sandstone leached out iron in some places, and re-deposited it in other places. When the groundwater was supersaturated with iron, the iron was precipitated around some kind of nucleus, forming the concretion. In someplaces there are so many concretions that the sandstone looks like it has "measles," such as in Gateway Canyon north of the Calico Basin (Figure 49). There are places where there are so many weathered out marbles that they are like ball bearings and a little dangerous to walk on!

Figure 48- Iron concretions in the Aztec Sandstone in the Calico Hills. They were formed by precipitation of iron by groundwater. The iron-impregnated concretions are more resistant to weathering and erosion than the surrounding sandstone. Diameter of the balls is up to about one inch.

Figure 49 - Iron concretions in the Aztec Sandstone in Gateway Canyon, north of the Calico Basin. The red concretions, some over one inch in diameter, make the rock look like it has "measles." Photo by David Morrow.

Discovery of Dinosaur Footprints in Red Rock Canyon

Six dinosaur tracksites have been found in the Jurassic rocks in Red Rock Canyon. Two of the sites are near the Calico Basin and one is in the Calico Hills. They are difficult to reach, and two are on dangerous cliffs. One site is near Willow Springs and is difficult to reach. Two sites are near Pine Creek Canyon.

Most of the tracksites are on a rock shelf or ledge that is covered, and protected from erosion, by desert varnish.

BLM guidelines do not allow revealing the exact location of these places in order to protect them from possible vandalism or theft. Photographs of the dinosaur footprints at two of the sites are displayed on a poster near the picture window in the Visitor Center in Red Rock Canyon National Conservation Area.

The six dinosaur tracksites are discussed below.

Dinosaur Tracksites #1 and #2

Hikers Discover the First Dinosaur Tracks (Site #1)

In April of 2010, three avid hikers – Gary Smith, Lynn Nicholson, and Jeff Mishlove came upon some unusual features on top of a cliff in the Jurassic Aztec Sandstone. Smith, retired from the FBI, suspected that they were tracks, possibly dinosaur tracks (Figure 50). Most of the tracks were smudgy and looked as though the animal or animals stepped on top of its own footprints, so it was not obvious that these were dinosaur footprints. Nicholson photographed some of the tracks and brought the photos in to the BLM. Although the photographs were excellent, the tracks that were photographed did not show a clear tridactlyl (three-toed) print and were inconclusive.

A Geologist Arrives at Red Rock Canyon

In March 2011, I was hired by the Red Rock Canyon Interpretive Association (now the Southern Nevada Conservancy). The SNC assists the BLM by providing interpretive expertise by way of hikes and talks, and also runs the gift shop and fee booths. In August 2011 I was giving geology training to the BLM volunteers. I mentioned that we were surrounded by Jurassic rocks, and it was only a matter of time until we found dinosaur footprints. (I expected to find them in the Kayenta-Moenave Formation below the Aztec Sandstone. This is a prolific dinosaur fossil unit in Southwestern Utah. The climate was humid during Kayenta-Moenave time with lakes and swamps and rivers and floodplains along which dinosaurs lived, at least in Utah.)

Gary Smith, who was in the audience, said "I think we already found some." When he said it was in the Aztec Sandstone I was skeptical since the sandstone was deposited as sand dunes during Jurassic time. Sand dunes are not a good depositional environment for preserving fossils, including footprints, and do not support animals as well as other environments such as lakes and swamps. Although dinosaur footprints had been found in the age-equivalent Navajo Sandstone in Utah and elsewhere in the region, no dinosaur footprints had been reported in Nevada.

Confirmation of the Dinosaur Tracks by the Geologist

On a hot August morning two weeks later, Smith and Nicholson led me on a hike up to their site. We started a little later in the morning than we should have, and the three of us almost died up there. (See "Dinosaurs Don't Give Up Their Secrets Easily," below). At the Smith track site, on a dangerous cliff, I confirmed that the hikers did indeed find what appeared to be dinosaur footprints (Figure 50). I would have to call in a paleontologist to verify and identify the tracks. The tracks were protected by a layer of desert varnish. Desert varnish is comprised of manganese and iron oxides that form on siliceous rocks such as sandstone and serves to prevent weathering and erosion of the rock surface (also see section on desert varnish, below).

Geological Treasures of Red Rock Canyon NCA and Spring Mountain Ranch S.P.

Figure 50 - Dinosaur tracks at the Smith site, the first dinosaur tracksite found in Red Rock Canyon, in the Jurassic Aztec Sandstone near the Calico Basin. The two tracks indicated show the tridactyl (three-toed) form. They are about five inches long.

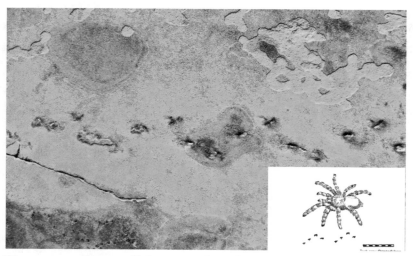

Figure 51 - *Octopodichnus* tracks. These were made by an arthropod - a spider or scorpion - that lived in the Jurassic sand dunes. *Octopodichnus* is the name of the track, not the name of the animal that made the track. Each footprint is about 1/4 inch long. Photo by Lynn Nicholson.

Discovery and Origin of the Iconic Dinosaur Track (Site # 2)

On the way up to the Smith site, we found a solitary dinosaur footprint. As always, I inspected every bedding plane within reach while hiking up to the site, looking for tracks. About 700 feet from the Smith site I found small animal tracks - a scorpion or spider (Figure 51). Nearby was a beautifully preserved dinosaur track (Figure 52). Nicholson took a photograph of the dinosaur track which was published in the *Las Vegas Review Journal* a few months later when the BLM was ready to reveal that "Dinosaurs Hiked Red Rock Canyon" (Figure 53). Notice the bulge and the rings around the track – either pressure ridges or collapse features in the wet sand, preserved after 180 million years.

Here is how the footprints are believed to have been preserved: The dinosaur stepped in moist sand probably after a rain, making a footprint, the sun came out and the tracks dried out and hardened, they were covered by blowing sand and added to the geological record. After 180 million years the layers with the footprints were exposed by erosion of the overlying layers. Desert varnish formed on the surface of the layer with the tracks, helping to preserve them.

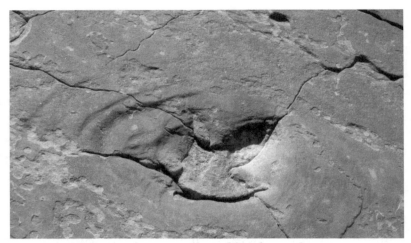

Figure 52 - Dinosaur footprint found by the author. Later identified by Professor Steve Rowland as a *Grallator* track. (This is the name of the track, not the name of the animal that made the track.) Notice the bulge as the dinosaur stepped into moist sand 180 million years ago.

Figure 53 - Newspaper article by Henry Brean on the discovery of dinosaur footprints in Red Rock Canyon which appeared in the *Las Vegas Review Journal* on November 21, 2011. The photo of the dinosaur footprint was taken by Lynn Nicholson.

Identification of the Tracks by a Paleontologist

Smith and I returned to the two track sites in early September 2011 with paleontology Professor Steve Rowland of the University of Nevada, Las Vegas (UNLV). Professor Rowland identified the tracks as ichnogenus (trace fossil genus) *Grallator*. *Grallator* is the name of the track, not the name of dinosaur that made the track. Several different species of dinosaurs could make *Grallator* tracks. Finding body fossils (bones or teeth) is necessary in order to identify the species of dinosaur. However, based on similar findings in rocks of the same age in other states, the dinosaur species is believed to be *Coelophysis* or *Megapnosaurus*. These are bipedal (walked on two feet), tridactlyl (three-toed) therapods (meat-eaters). The track makers at RRCNCA were the size of a large dog. All of the dinosaur tracks in Red Rock Canyon are *Grallator* tracks. Figures 54 and 55 are drawings of what the dinosaur probably looked like. Figure 55 also shows a drawing of an early mammal.

Professor Rowland identified the small animal tracks as ichnogenus *Octopodichnus*. These were made by arthropods such as scorpions or spiders about the same size as those that roam the sand dunes today. (See Figure 51.)

Figure 54- Drawing of dinosaur that made the *Grallator* footprints. May be a *Coelophysis* or *Megapnosaurus*. Note the tridactlyl (three-toed) tracks. Source: "Desert Dinosaurs"

Figure 55 - Drawing of desert dinosaurs in the sand dunes. Early mammals were dinosaur food, and one of dinosaurs got his lunch. Source: ResearchGate.com; image by Ariel Milani

Verification and Mapping by BLM Paleontologists

The discoveries were reported to the BLM. In late October 2011 the BLM organized an expedition to the sites, guided by Smith, Saines, and Rowland. The BLM team consisted of Regional Paleontologist and track expert Brent Breithaupt, Regional Paleontologist Scott Foss, cartographer Neffra Matthews, and Park Ranger Robert Valenzuela. The paleontologists verified the discoveries, and assisted Matthews with three-dimensional imaging photography and mapping of the tracks. Near the *Octopodichnus* tracks Valenzuela discovered another small animal trackway, which has been tentatively identified by the paleontologists as proto- or early mammal.

Dinosaur Tracksite #3
Near Pine Creek Canyon

In March 2012, Sendi Kalcic, BLM Wilderness Specialist in Las Vegas, found a large block of rock with dinosaur and small animal trackways near Pine Creek, within the Rainbow Mountain Wilderness (Figure 56). This is the best dinosaur tracksite in Red Rock Canyon. Kalcic was working with Washington State University students who were volunteering as part of their Outdoor

Figure 56 - *Grallator* dinosaur tracks alongside *Brasilichnium* early mammal tracks in the Jurassic Aztec Sandstone near Pine Creek Canyon. This is the best trackway in Red Rock Canyon. (*Grallator* and *Brasilichnium* are the name of the tracks, not the name of the animals that made the track.)
Photo by Sendi Kalcic, formerly BLM.

Program. They were posting wilderness signs at the base of the great cliffs of Jurassic Aztec Sandstone in Red Rock Canyon. She noticed and photographed sets of tracks on a large block of rock that had fallen from the cliffs above. She submitted the photographs, and experts reviewing them could see that it was a genuine and important find. In addition to the dinosaur footprints, the slab shows what has been identified as proto- or early mammal tracks alongside the dinosaur tracks. The tracks, not the track-maker, have been identified as *Brasilichnium,* by Professor Rowland. The tracks are protected by a thick coating of desert varnish.

Unlike the other three track sites in the Calico Hills, which are very difficult to reach and, for the most part, are on dangerous cliffs, this site is accessible by a strenuous one hour hike. It is difficult to climb up onto the block itself, but the tracks can be viewed from adjoining high ground without the need to climb the rock. Therefore should the BLM decide to open a track site to the public this would be the best candidate, as it has the best trackways, it is the most accessible, and the tracks can be best viewed from adjoining high ground rather than trying to climb onto the rock itself.

Dinosaur Tracksite #4
The Calico Hills Site

In October 2012, Aaron Leifheit, who was naturalist with the Southern Nevada Conservancy, hiked to a difficult to reach ledge or shelf in the Calico Hills. Along the route and on a cliff overlooking Red Rock Canyon, he found dinosaur footprints and numerous small animal tracks. The small animal tracks include *Octopodichnus* or arachnid (spider or scorpion) and proto-mammal tracks. Two sets of tracks at the site are shown in Figure 57. Shelves or ledges may sometimes represent breaks in Jurassic sand dune deposition such as an interdune area, where preservation of tracks is more likely.

Dinosaur Tracksite # 5
Pine Creek Canyon

In the summer of 2015, Rob Shaw discovered a boulder of Aztec Sandstone containing a dinosaur track while hiking in Pine Creek Canyon. As in most places, the track has been protected from erosion by the desert varnish. On this boulder most of the desert varnish has been eroded away, but a little remains in the dinosaur track (Figure 58). The print appears to be a *Grallator* track. This is the name of the track, not the name of the dinosaur.

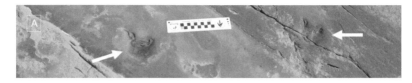
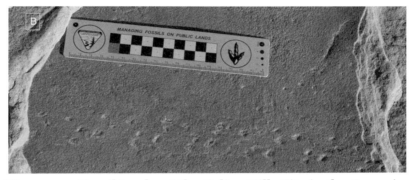

Figure 57 - Fossil tracks at the Calico Hills Site in the Jurassic Aztec Sandstone.
A - *Grallator* dinosaur tracks
B - *Octopodichnus* tracks (spider or scorpion).
Checkered scale is 10 cm (4 inches) long.

Figure 58 - Dinosaur footprint (*Grallator*) on a boulder in Pine Creek Canyon. Desert varnish has preserved the track.

Dinosaur Tracksite # 6
The Willow Springs Site

High on a cliff in a canyon that drains into Red Rock Wash near Willow Springs, hiker J. Michael Tracy discovered dinosaur footprints in October 2016. The prints are on a ledge of Aztec Sandstone protected by a thin veneer of desert varnish (Figure 59). There are also tiny tracks at the site made by an unknown animal, perhaps an arthropod.

Figure 59. Dinosaur trackway near Willow Springs in Jurassic Aztec Sandstone. Inset shows enlargement of top two tracks. Each colored bar is 10 cm (4 in) long.

"Dinosaurs Don't Give Up Their Secrets Easily"

All three "dinosaur hunters" almost didn't survive the August, 2011 hike to the dinosaur footprints (see previous section). After the Saines site was photographed, the three were examining and photographing the Smith site, 700 feet to the east, located on a dangerous cliff. I was so excited to see the prints that I was backing up and taking pictures without thinking about the certain-death drop-off. Gary Smith said, "Nick, you are getting me nervous." It saved my life - I was a few feet from the edge.

By this time it was mid-morning and getting very hot. I looked at the route we took to get to the Saines and Smith sites, and it entailed over a thousand foot long up-hill climb through the rocky terrain. We decided to try to find a route directly down instead of going back the way we came. Then Gary Smith's cell phone rang – it was his wife. Getting cell phone service is pretty spotty in Red Rock Canyon, but this time it worked. She said her car wouldn't start, and she needed to get to work. Lynn Nicholson volunteered to go back to where we parked the cars, but he tried to take a short cut, not the way we came.

Gary and I headed down. It was late August and late morning, and it was well over 100° F. The heat was radiating from the rocks like an oven. I had plenty of water, but it was warm and my core body temperature was going up. I couldn't cool down, and I started to feel dizzy. Gary was also feeling the heat. We worked our way down a steep rocky ravine towards the valley floor, stopping under every bush or tree that afforded some shade to help cool us down. Finally we made it to the valley floor. We had a ¼ mile walk to a parking lot and shade. We were going to hitch a ride to the Visitor Center and then get a ride to Gary's car. I was so wobbly Gary had to hold me up a few times as we hiked to the parking lot.

When we arrived at the parking lot there were Search and Rescue vehicles there. We assumed that someone saw us struggling and called for them to come to rescue us. They gave us cold water, cold towels for our neck, and had us sit in the shade. In a few minutes we felt much better. One of the rescuers said, "Which one of you has a broken foot?" We shrugged. They said that they got a call that someone broke their foot on the mountain behind us. It had to be Lynn! We pointed in the general direction he went, but we were in no condition to go back up the ridge and show them where he might be.

Gary got through to his wife on his cell phone. Although it was a bad connection he was able to converse with her. He said to the rescue personnel, "My wife got a call from Lynn. He broke his ankle, but he has been rescued." But how could that be? Everybody was mystified. Just then we heard, then saw, a helicopter circling above the ridge. Lynn had told her that he was *going to be* rescued, not has been rescued. Lynn was lucky that two hikers with cell phones that worked found him in that remote location and were able to call 911.

So all three of us could have died out there that day. – *Dinosaurs don't give up their secrets easily*.

MESOZOIC ERA - CRETACEOUS PERIOD

Conglomerate of Brownstone Basin

The only Cretaceous formation in Red Rock Canyon is the Conglomerate of Brownstone Basin. Brownstone Canyon is the next drainage to the north of the Scenic Drive drainage basin. Because of the magnificent exposures of Jurassic sandstones, Red Rock Canyon has been called "The Real Jurassic Park." There are also good exposures of Triassic rocks – the Moenkopi and Chinle Formations. But the Cretaceous is a thin (less than six feet thick) conglomerate found in Brownstone Canyon, and in a few places along the thrust fault system, above the Jurassic Aztec Sandstone and below the fault breccia of the Cambrian Bonanza King limestone. It is found at the north end of the fenster or window at the end of the Keystone Thrust Trail (see Figure 88).

Figure 60 shows the formation in Brownstone Canyon. The conglomerate contains clasts or pebbles and cobbles of Mesozoic and Paleozoic rocks, including quartzite. It was deposited on an erosion surface cut into the upper part of the sandstone before the thrust faulting.

The origin of this formation is uplift to the west during Cretaceous time and erosion of the underlying Mesozoic and Paleozoic rocks. The rivers carried sediments towards the sea, which was located to the northeast in Lower Cretaceous time, according to Blakey's paleogeographic map (Figure 61).

The Cretaceous Conglomerate of Brownstone Basin correlates with the Cretaceous Willow Tank Formation, which is a conglomerate in its lower or basal part, in the Valley of Fire, 50 miles northeast of Red Rock Canyon. In the Valley of Fire there are thick Cretaceous formations that contain dinosaur bones and teeth. It is possible, but unlikely, that fossils would be found in the formation in Red Rock Canyon, which represents river channel deposits.

Figure 60 - Cretaceous Conglomerate of Brownstone Basin contact with underlying Jurassic Aztec Sandstone in Brownstone Canyon. Bill McKinnis to right. Inset shows an enlargement of the contact.

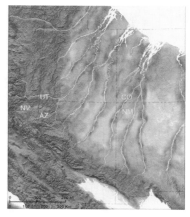

Figure 61 - Lower Cretaceous paleogeographic map. Notice drainage northeasterly from southern Nevada.
Source: Blakey and Ranney (2008)

CENOZOIC ERA

Ancient Landslides

During the late Paleogene or Early Neogene Periods, about 25 million years ago, giant landslides came crashing down from the limestone that was precariously over-thrust on top of the sandstone along the Wilson Cliffs escarpment. The great mass of crushed and broken limestone was unstable, and the landslides may have been triggered by seismic activity (earthquakes) or just gravity collapse. According to Page and others (1998) this may have occurred when the overthrust plate was close to its maximum topographic relief. Processes of gravitational collapse are common in thrust belts where the overthrust plate becomes unstable, collapses, and slides laterally forming landslide breccia.

One landslide occurred in the vicinity of Blue Diamond, and one occurred just north of the Calico Basin. The geologic map (Figure 3) shows these as Tertiary landslides. (The Paleogene and Neogene Periods were formerly lumped together as the Tertiary Period.) These are long runout landslides or *sturzstroms* that killed everything in their paths, although no Cenozoic fossils have been found in the landslide deposits. The landslide deposits, unlike the younger debris flow deposits (see below), are consolidated rock. The deposits are *landslide breccias* – mostly crushed and broken limestone fragments cemented together. But there are also pieces of sandstone included in places.

Blue Diamond Landslide

As shown on the geologic map (Figure 3), the landslide breccia caps Mesozoic rock ridges in the vicinity of Blue Diamond (Figure 62). One large block of sandstone that was swept up in the landslide is found on a ridge just west of Blue Diamond (Figure 63). In the photos and captions the formations are labeled as Tertiary rather than late Paleogene or early Neogene in conformance with the geologic map.

There is a deposit of the landslide breccia on top of Blue Diamond Hill in the gypsum mining area. This means that the landslide runout was about five miles from the escarpment, which, when the sturzstrom occurred, was farther east and higher than it is now. This occurred when the bottom of Red Rock Canyon east of the escarpment was not as deep below the top of the escarpment as it is today. Millions of years of erosion have deepened the valley through which Route 159 now runs.

Figure 62 - View of village of Blue Diamond, looking west showing Tertiary landslide deposits capping the ridge behind the village.

Figure 63 - Sandstone block incorporated in Tertiary landslide in Blue Diamond. Note people for scale on top of block in inset.

Turtlehead Landslide

The Turtlehead Landslide is located north of the Calico Basin (Figure 64, and geological map, Figure 3). According to Peck (2000) the landslide off the thrust plate infilled a valley. The cemented landslide breccia was more resistant to erosion than the bedrock walls of the valley so that there is a reversal of topography with the former valley now being a ridge.

Some of the breccia clasts (fragments) are meters thick and retain the original bedding and fossils, but they became jumbled up during the landslide.

In Brownstone Canyon there is an unusual outcrop of the limestone breccia with vertical rills due to dissolution of the carbonate over time as water runs off the outcrop (Figure 65). A hiker remarked that this outcrop would be kind of spooky looking on a dark and misty evening. I concur. Maybe we should take a Halloween evening hike there to find out!

Figure 64 - Tertiary Turtlehead Landslide, looking north from Calico Basin.

Figure 65 - Limestone landslide breccia with vertical rills in the Turtlehead Landslide in Brownstone Canyon.

Potato Knoll – A Slump Block

Potato Knoll (Figures 26, 66, and 73) is a slump block that broke off from the Wilson Cliffs, probably in the late Pleistocene or Holocene (Recent), and rotated out on the mudstones of the Petrified Forest Member of the Chinle Formation. Figure 67 is a sketch cross section. The Shinarump Conglomerate, including the sandstone layer above the conglomerate, is in place below the Petrified Forest Member, and the huge block of rock slid on a weak clayey shale layer above the Shinarump. This giant slump block is the only such feature along the escarpment. Clay layers in the Petrified Forest Member are a common cause of landslides in southern Utah, including in and near Cedar City, and in Zion National Park. The slump block included Kayenta-Moenave shales below the Aztec Sandstone that have been eroded away, except for the northeast corner of Potato Knoll, where they are still exposed.

There is a large hole or alcove in the cliff face near the top of Mt. Wilson (Figure 66). It has also been hypothesized that Potato Knoll fell thousands of feet from what is now the alcove to the valley bottom. However, geologist John Peck and the author believe that since the sandstone is intact at the top of Potato Knoll, a slump along a clay layer is more likely. A rock fall of thousands of feet would be expected to smash the block to pieces.

Figure 66 - Mt. Wilson, showing Potato Knoll near its base and the alcove near the top.

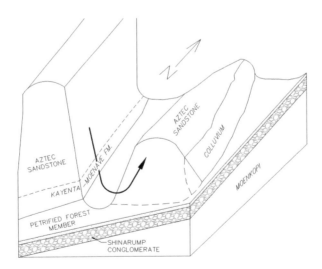

Figure 67 - Sketch cross section through Potato Knoll. A small remnant of the red Kayenta shale is exposed on the northeast corner of Potato Knoll (not shown on the cross section).
Source: Modified from cross section by John Peck.

Recent Debris Flows

Debris flow boulder deposits are found in all the canyons along the base of the escarpment, including Juniper Canyon, Oak Creek, Pine Creek, and First Creek. A debris flow is a type of landslide. It is a downward mass movement of loose material – a watery mass of sand, gravel, cobbles, and boulders that can travel more than 25 mi/hr down a canyon and onto the alluvial fan. Debris flows are typically triggered by severe rain storms. Remnants of debris flows are recognizable by the *boulder levees* which form as the main watery thrust of the debris flow carves out a channel, leaving boulders on either side. An excellent example of debris flow boulder levees is found in Oak Creek Canyon north of Potato Knoll (Figure 68). Downstream of Juniper Canyon a debris flow deposit is found with a boulder levee over 10 feet high (Figure 69). The age of these unconsolidated deposits is believed to be Recent or Holocene (less than 10,000 years old).

Can a debris flow come roaring out of one of the canyons today, destroying everything in its path? – Yes, it is possible, even though a lot of loose material has already been removed by older debris flows. (Also see "Natural Dam in Juniper Canyon," below.)

Figure 68 - Debris flow boulder levees, Oak Creek Canyon, north of Potato Knoll, looking south. With Vern Quever (left) and Sasson Jahan (right).

Figure 69 - Debris flow boulder levee downstream of Juniper Canyon.

Natural Dam in Juniper Canyon

The most unusual canyon in the Wilson Cliffs escarpment of Red Rock Canyon is Juniper Canyon, located south of Pine Creek Canyon (Figure 70). Unlike Pine Creek Canyon and Oak Creek Canyon, Juniper Canyon doesn't have a mouth. The mouth is blocked by what appears to be a huge landslide comprised mainly of large blocks of sandstone mixed with sandy soil (Figures 71 and 72). It is not known if there is a bedrock lip or ledge covered by the natural dam.

There is a drainage cut into the dam along the north bedrock wall. The only way into the canyon is by climbing up the steep dam face, about 200 feet high. Not an easy task. Once on top of the dam you descend down into the canyon bottom, which is hundreds of feet higher than the bottom of Pine Creek.

The natural dam probably formed since the end of the last Ice Age, less than 12,000 years ago, and may have been triggered by an earthquake.

An extremely high rainfall event could cause ponding behind the natural dam in Juniper Canyon. This could potentially be the set up for a future debris flow out of the canyon. Fortunately there is no infrastructure (roads, parking areas, rest rooms, signage) directly downstream of Juniper Canyon. The Scenic Drive is about 1.5 miles east of the canyon.

Juniper Canyon is reachable by about a one hour hike along the Arnight Trail to the Knolls Trail from Oak Creek Canyon, or from Pine Creek, connecting up with the Knolls Trail.

Figure 70 - Juniper Canyon is hundreds of feet higher than the mouths of Oak Creek and Pine Creek canyons due to a natural dam, probably a landslide, blocking its mouth.

Figure 71 - Mouth of Juniper Canyon blocked by a natural dam of sand and blocks of sandstone. With (from left to right) Gareth Pearson, Pete Stephenson, Tom Lisby, and Gary Smith.

Figure 72 - Climbing the natural dam in Juniper Canyon, comprised of sand and blocks of sandstone. With (from left to right) Pete Stephenson and Tom Lisby.

Skull Rock Landslide North of Pine Creek Canyon

Less than ½ mile north of Pine Creek Canyon on Dale's Trail is terrain with large displaced Aztec Sandstone blocks, including Skull Rock (Figures 73 and 74). This is interpreted as a Holocene landslide which came down from the face of the escarpment about 1000 feet to the west. This may have occurred in several pulses over time. There are other similar landslides along the escarpment. There are also rockfalls at the base of the escarpment, where rocks have fallen off the cliffs.

Figure 73 - Skull Rock landslide, looking south from Dale's Trail. Photo by Jeff Cuneo.

Figure 74 - Skull Rock, looking west towards escarpment. Red sandstone block probably came down from red unit in Aztec Sandstone thousands of feet above the valley floor. Skull Rock is about 15 feet high.

Precariously Balanced Rocks

Precariously balanced rocks (PBRs) are rocks that look like they would easily be toppled in an earthquake. If an area has a lot of PBRs, one can assume that the area is not seismically active, because strong earthquakes would be expected to have toppled them. Red Rock Canyon has numerous PBRs and is not in a seismically active area. Seismologist Jerry King and the author presented a paper at the Fall 2013 national meeting of the Association of Environmental and Engineering Geologists about seismic activity and PBRs in Red Rock Canyon (see References).

PBRs can be attached to the bedrock below or detached blocks of rock. Some PBRs are boulders or blocks of rock that have become detached from the bedrock and are perched precariously at the edge of a cliff. Figure 75 shows Mushroom Rock and the wine glass terminology proposed for bedrock PBRs. Mushroom Rock is located near the west end of the Calico Hills Trail, between Calico II and the Sandstone Quarry parking lot. The narrower the stem the more precarious the rock. Another excellent example is in Brownstone Canyon (Figure 76). Figure 77 shows a PBR at the East Point on the Calico Hills Trail. This is a rock favored by climbers that gets hikers nervous. It looks as if the climbers could pull the rock over on top of them.

Figure 75- Mushroom Rock near west end of Calico Hills Trail, a precariously balanced rock (PBR), labeled with wine glass terminology. The stem is also known as a pedestal in some geologic reports on PBRs.

Figure 76 - Precariously balanced rock in Brownstone Canyon. Photo by Harold Larson.

Figure 77 - Precariously balanced rock at the East Point on the Calico Hills Trail. This rock is favored by rock climbers who should look at this side view before attempting to climb it.

Spring Deposits in the Calico Basin

Along the western edge of the Calico Basin between Red Spring and Ash Creek Canyon, spring deposits are found (Figures 78 and 79). The deposits, north of the boardwalk at Red Spring, are several feet thick, and are composed mainly of tufa (a form of limestone deposited by springs), and limestone or tufa conglomerate. The limestone conglomerate has clasts or pieces of red sandstone incorporated in a limy matrix deposited by the springs. Some of the deposits may be degraded spring mounds. The underlying bedrock is the Aztec Sandstone.

These spring deposits are probably Pleistocene (Ice Age) in age, when the climate was wetter and cooler. There is north-south faulting along the west side of the Calico Basin. The springs are believed to be related to upward discharge or flow along the fault.

Another hypothesis is that the spring deposits are similar in age and origin to those in Tule Springs Fossil Beds National Monument in North Las Vegas. At Tule Springs the deposits are interpreted as being formed in wetlands. It may be that the Calico Basin was a wetland or lake at times during the Pleistocene, and the springs discharged into the wetland. Since the time of deposition, erosion has left the deposits high and dry. All that remains of the spring activity are the small springs nearby at a lower elevation. These are Red Spring, Calico Spring, and, in Ash Creek Canyon, Ash Creek Spring. These springs drain the east side of the Calico Hills. During the Pleistocene, especially during the Glacial Maximum, about 25,000 years ago, the precipitation, and therefore spring discharge, was much greater.

Figure 78 - Spring deposits on the west side of the Calico Basin below the base of the sandstone cliffs.

Figure 79 - Close up of spring deposits - tufa and tufa conglomerate with incorporated pieces of sandstone.

Desert Varnish - "The Rocks Are Alive"

The black patina visible on many sandstone cliff faces (vertical or near vertical) and bedding planes (relatively flat or gently-dipping rock surfaces) is called desert varnish (Figure 80). Desert varnish is composed of manganese and iron oxide that is fixed by bacteria. The latest theory of its origin is that it is developed in arid regions on clay-sized dust particles blown onto the rock face by the wind, and acted on by manganese- and iron-oxidizing microbes or bacteria. It forms on sandstone rather than limestone, because the high alkalinity of limestone minerals and the soluble surfaces are not conducive to its formation. Desert varnish is believed to take many hundreds or thousands of years to form.

Native Americans carved petroglyphs into the desert varnish in Red Rock Canyon and in many other places in the Southwest. (See section on petroglyphs below.) Some of the best exposures of dinosaur footprints and trackways are covered and protected by a patina of desert varnish (e.g. Figure 56).

Some petroglyphs are believed to be thousands of years old, as they have developed desert varnish on them - almost totally obscuring the petroglyphs in some places (Figure 81).

Since the formation of desert varnish is believed to

be on-going today, one might say that "the rocks are alive," this phenomenon has been referred to as a "shadow biosphere" by biologists.

Figure 80 - Desert varnish on ripple-marked bedding plane surface in the Aztec Sandstone in the Calico Basin. Wind ripples formed on the windward side of a sand dune during Jurassic time. Desert varnish helps protect the surface from erosion.

Figure 81 - Desert varnish developed on very old petroglyphs, almost totally obscuring them. East Point, Calico Hills.

RED ROCK CANYON FAULTS

Faults are places where the tectonic forces in the Earth caused rocks to break and move along the fault plane. Faults are common throughout southern Nevada, as in many other places, and there are four major faults in Red Rock Canyon: the Keystone Thrust Fault, the LaMadre Fault, the Red Spring Thrust Fault, and the Cottonwood Fault (see the geologic map, Figure 3) The "fenster" or window eroded through the Red Spring Thrust Fault is also discussed.

The Keystone Thrust Fault - Red Rock Canyon's Famous Fault

The most dramatic geologic feature of Red Rock Canyon, after the great Wilson Cliffs sandstone escarpment, is the Keystone Thrust Fault. Looking west from the viewing platform at the Visitor Center or anywhere on the east side of the park, and in Spring Mountain Ranch, one can see gray rock sitting on top of red and buff sandstone cliffs (Figures 37 and 42). The gray rock is Paleozoic limestone. At its base is the Bonanza King Formation, a limestone over 500 million years old. It is sitting on top of 180-190 million year-old Jurassic rock, the Aztec Sandstone.

Normally in geology, younger rock lies on top of older rock. Here we have older rock on top of younger rock. – How did this happen?

During the late Mesozoic Era the eastward-moving oceanic Pacific Plate began subducting (moving downward) below the westward-moving North American plate, causing compression of the upper layers of the crustal rock and thrusting of Paleozoic formations to the east. Also, the subduction caused rocks to melt at depth, and the formation of volcanoes in Southern California. Figure 82 is a block diagram from Tingley and others (2001) that illustrates the process. The great granite batholith (giant mass of igneous rock) of the Sierra Nevada began to form, and it may have contributed to the displacement and thrusting of the older Paleozoic carbonates eastward over the Jurassic sandstone.

According to Burchfield and others (1974) the rocks were thrust as much as 23 km (13.8 miles) to the east. If the average rate of movement was roughly two inches a year (based on the rate of movement of some faults in California), doing the math, it took more than 400,000 years for the thrust plate to reach what is now Red Rock Canyon. The rocks have been eroded back thousands of feet to the west since the end of the thrust faulting.

This regional thrust faulting, of which the Keystone Thrust Fault is a part, is called the Sevier Orogeny, which ended in the Cenozoic Era about 50 million years ago. The Keystone Thrust Fault system can

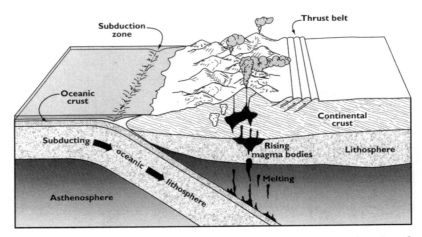

Figure 82 - Block diagram from Tingley and others (2001) showing the oceanic plate subducting (moving down below) the continental plate causing compression and thrust faulting in the continental crust during late Mesozoic time.

be traced north into the Valley of Fire, through Utah, into the Rocky Mountain states and into Canada. But arguably the best exposure is right here in Red Rock Canyon!

Burchfiel and others (2010) differentiate between the Keystone thrust plate and the Wilson Cliffs thrust plate with the latter being structurally below the former. In this book the plates are undifferentiated, and, along with the Red Spring Thrust Fault, are all considered part of the Keystone Fault Thrust System.

Figures 83 and 84 show how the fault dragged up

the formations as it was thrust over them on the west side of the White Rock Loop.

Figure 83 - West side of White Rock Loop looking north showing steeply dipping rock layers that were dragged up by the Keystone Thrust fault. Limestone to west was once over the top of these layers, and has been eroded back over the past 65 million years.

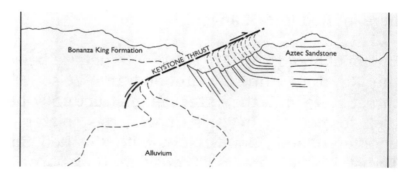

Figure 84 - Diagram from Tingley and others (2001) that shows how the rock layers were dragged up by the Keystone Thrust Fault on the west side of the White Rock Loop. Same view as Figure 83 above. Also see Figure 27.

La Madre Fault and a Reversal of Topography

Another important fault and major feature of the Red Rock Canyon landscape is the La Madre Fault. The near-vertical La Madre Fault trends northwest-southeast across the northern part of the National Conservation Area, and can be traced for more than 30 miles across the Spring Mountains. See Figure 22.

The fault can be seen from the Visitor Center viewing platform as a fracture or crack in the La Madre Mountain limestone cliffs just northwest of the north end of the White Rock Hills (Figure 85).

The fault is responsible for the abrupt scarp of the Calico Hills along the Scenic Drive (Figures 7 and 86). Here younger Jurassic rock on the north side of the fault was down-dropped alongside older Permian rock, yet it is higher in elevation than the south side. This is called a *reversal of topography.* On the south side of the fault, south of the road, there are outcrops of the Permian Kaibab Formation, which has been eroded down more than the Jurassic sandstone on the north side of the fault. Also see Figure 22.

Figure 85 - The trace of the La Madre Fault is visible cutting through the limestone in the overthrust plate.

Figure 86 - The abrupt scarp of the Calico Hills was formed by movement along the La Madre Fault. The red and buff-colored Jurassic sandstone moved down on the north side of the fault, yet it is higher than the south side. This is known as a *reversal of topography*.

The Red Spring Thrust Fault

The Red Spring - Calico Hills Thrust Fault is part of the Keystone Thrust Fault System in which Paleozoic limestones were thrust on top of Jurassic Sandstones. The fault is beautifully exposed in the northwest part of the Calico Basin where the limestone/sandstone fault contact is clearly seen (Figure 87). The limestone is the Cambrian Bonanza King Formation, over 500 million years old. It overlies the 180-190 million year old Jurassic Aztec Sandstone. The contact can be seen by hiking up "Five Stop Hill," where limestone fault breccia is found overlying the sandstone. The Red Spring Fault continues to the west along the north side of the Calico Hills and is found in the fenster described below.

Figure 87 - Red Spring Thrust Fault in Calico Basin looking northwest. Gray Cambrian Bonanza King limestone thrust over red and buff-colored Jurassic Aztec Sandstone.

The Fenster - A Window Through the Red Springs Thrust Sheet

The Keystone Thrust Trail, Trail No. 5 off of the Scenic Loop Drive, takes you to a geological phenomenon called a "fenster" (the German word for "window") (Figure 88). The fenster is a window eroded through the overthrust sheet or plate of the Red Spring Thrust Fault. As discussed above, a thrust fault is a fault in which one plate rides over the other, as opposed to down-dropping or moving laterally along the fault.

The Red Spring Thrust Fault is part of the Keystone Thrust Fault system. The overthrust plate is Cambrian limestone, and it was thrust over the Jurassic sandstone and a thin layer of Cretaceous conglomerate, known as the Conglomerate of Brownstone Basin (see separate section on this formation, above). On the north side of the fenster, the rock in contact with the sandstone and conglomerate is limestone fault breccia – crushed and broken rock due to the faulting (Figure 89).

The fenster (Figure 88) is a beautiful exposure of the colorful sandstone that lies beneath the overthrust plate. It has been exposed by a wash that cuts through the overlying limestone revealing the red sandstone below. The red color is due to the mineral hematite, an iron oxide. Other sandstone colors that are found in the fenster are brown, yellow, and rust due to iron oxide minerals limonite and goethite, and white due to the leaching or removal of iron by groundwater seepage. (Also see the section on Jurassic rock colors, above.)

To reach the fenster, drive to the White Rock Spring parking lot and hike the Keystone Thrust Trail for a little over a mile.

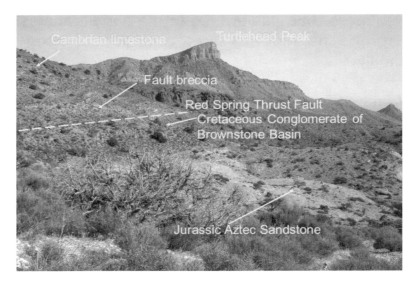

Figure 88 - The fenster or window at the end of the Keystone Thrust Fault Trail, looking east. The gray Cambrian limestone which was thrust over the sandstone has been eroded away revealing the colorful Jurassic sandstone below.

Cottonwood Fault - See it on the Road to Pahrump (Route 160)

The Cottonwood Fault, like the La Madre Fault, is a spectacular east-west striking (trending) vertical fault. Driving west on Route 160 towards Mountain Springs and Pahrump one notices that the red sandstone escarpment suddenly ends on the north side of road, and gray limestone is found at the same elevation as the sandstone on the south side of the road (Figure 90). The fault is a zone of weakness that was eroded into a saddle across the ridge. The highway was constructed along the saddle.*

Which side do you think moved down? With Paleozoic limestone opposite younger Jurassic sandstone one would expect that the younger sandstone moved down (as along the La Madre Fault). However, here, the block with the limestone was originally thrust on top of the sandstone and is believed to be the down-dropped block. If that is the case, if a hole was drilled through the limestone on the south side of Route 160, the Jurassic sandstone would be encountered at depth.

---

* It was along this fault, through the saddle, that Captain John C. Fremont, Kit Carson, and their men passed in 1844 along the Old Spanish Trail.

Figure 89 - Fault breccia in the Red Spring fault zone on the north side of the fenster. "Breccia" means "broken" in Italian - crushed and broken rock in the fault zone.

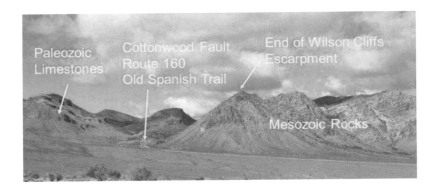

Figure 90 - Cottonwood Fault, looking west from Badger Pass, south of Route 160. The south end of the Wilson Cliffs Escarpment is at the Cottonwood Fault.

ARCHAEOLOGY - NATIVE AMERICAN ARTIFACTS

Native Americans reached the Las Vegas Valley at the end of the last Ice Age, roughly 10,000 years ago, on their journey from Asia across the Bering Land Bridge. Native Americans lived in the Valley and in Red Rock Canyon for thousands of years. Archaeological features left by the native Americans include petroglyphs, pictographs, agave roasting pits, and rock shelters.

Petroglyphs - Carved in the Rock

Petroglyphs are carvings in the rock made by the former Native American inhabitants. There are dozens of petroglyph sites in Red Rock Canyon. Although a few petroglyphs are carved directly into the rock, most petroglyphs are carved in desert varnish (the dark patina of manganese and iron oxide that covers the sandstone in places. See the discussion of desert varnish above).

Desert varnish takes hundreds to thousands of years to form. Some of the petroglyphs are so old that they have developed a patina of desert varnish on them, all but obscuring the petroglyphs (Figure 81). It could be that there are petroglyphs in Red Rock and elsewhere that are thousands of years old and have been completely obscured by desert varnish and are invisible to the naked eye. It is just a matter of time until someone invents a light, like a UV black light flashlight, that can detect and reveal petroglyphs thousands of years old that were completely covered by the younger patina.

Petroglyphs can be seen in rocks alongside the boardwalk at Red Spring (Figure 91), at the end of the Petroglyph Wall Trail in Willow Springs (Figure 92), and in Brownstone Canyon (Figure 93). Another location is at the East Point – the east end of the Calico Hills, where huge blocks of sandstone toppled from the cliffs. Several of these blocks have petroglyphs carved in the desert varnish, including some that are becoming obscured by desert varnish forming on the petroglyphs (Figure 81). Most of the petroglyphs in Red Rock Canyon appear to be abstract designs that cannot be readily interpreted, but some show recognizable animals, especially bighorn sheep (Figure 94).

Petroglyphs Related to The Old Spanish Trail

In 1829 there was a historical event that opened up this region, which was then part of Mexico, to non-native Americans and is reflected in petroglyphs: The Old Spanish Trail that linked Santa Fe with Los Angeles was blazed. The Old Spanish Trail was used from 1829 to 1848 and ended with the Mexican War in which the southwest became U.S. territory. In 1848 the

Figure 91 - Petroglyphs on rocks along the boardwalk near Red Spring. Petroglyphs in Figures 91 and 92 are abstract designs that cannot be definitively deciphered.
Photo A by Betty Gallifent.
Photo B by Clint Wharton

Figure 92 - Petroglyphs at the end of the Petroglyph Wall Trail in Willow Springs. The petroglyph in the lower lower left resembles some of the pictographs and petroglyphs in Brownstone Canyon (Figures 93, 97, and 98).

Mormons took the first wagons on the Old Spanish Trail, linking Salt Lake City and Los Angeles. This became known as the Mormon Road (not the Mormon Trail, which goes from Illinois to Salt Lake City).

The Old Spanish Trail went from Las Vegas Springs to Cottonwood Springs (Blue Diamond) and along present-day Route 160 through what is now Mountain Springs. A variant of the main route is through Red Rock Canyon. An informational kiosk on the Old Spanish Trail is located at the Red Rock Overlook on Route 159. Seeing Mexican cowboys or muleteers and mule caravans, and later wagons, was a huge cultural shock to the Native Americans after thousands of years of isolation. Figure 94 shows a petroglyph of a cowboy (circled) near the trail to Turtlehead Peak.

Also at that site is a controversial "Falling Man" petroglyph, similar to the famous one in Gold Butte National Monument (Figure 95). Do you think it is a falling man? Both petroglyphs are etched into surfaces at rocky cliffs. Could it be a man who has fallen and broken his back?

Cowboys and wagons are depicted in faint petroglyphs at Lone Grapevine Spring close to Route 160 (Figure 96). In Figure 96 B, which has been enhanced, the wagon etching is missing, apparently due to rock weathering.

Figure 93 - Petroglyph panel in Brownstone Canyon. It is located about 40 feet above land surface and is about 25 feet wide. Note similarity of pattern in Figure 92 at Willow Springs.

Figure 94 - Petroglyph panel near trail to Turtlehead Peak. Note bighorn sheep, and cowboy figure (circled). The cowboy petroglyph suggests the petroglyph was carved circa 1829 - 1848 when Southern Nevada was part of Mexico, and mule caravans first began crossing the region on the Old Spanish Trail. Rock slab is about three feet across.

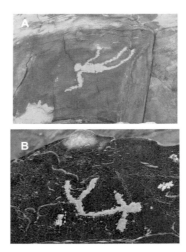

Figure 95 - A - Falling Man petroglyph in Gold Butte National Monument, Nevada. About one foot long.
B - Falling Man (?) petroglyph in Red Rock Canyon. Less than one foot long. Photo by Tom Lisby

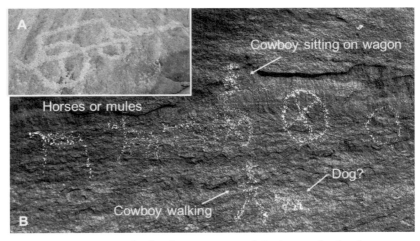

Figure 96 - A - Petroglyph of a wagon; and B - Petroglyph of a wagon with horses or mules and two cowboys. Both at Lone Grapevine Spring close to the Old Spanish Trail/Mormon Road (Route 160). Note dog (?) behind walking cowboy. Photograph B enhanced by Justin McAffee (Most Media). Petroglyphs are less than one foot long.

Pictographs - Paintings on the Rocks

Pictographs are paintings on the rocks made by Native Americans. They used ground up natural pigments, especially weathered iron oxide minerals (ochre) – hematite, limonite, goethite - to create red, yellow, and brown colors. Charcoal was used for black. Two places in Red Rock Canyon where pictographs can be found are Brownstone Canyon and Willow Springs.

Brownstone Canyon

To an archaeologist or anthropologist the pictographs in Brownstone Canyon are unarguably some of the most precious treasures of Red Rock Canyon. Native Americans, who lived in the Red Rock Canyon area for thousands of years, painted the pictographs on the rocks (and carved the petroglyphs into the rocks). What are probably the best pictograph panels in southern Nevada are in Brownstone Canyon. The panels contain interesting and elaborate and beautiful pictographs (Figures 97 and 98). No one really knows what they mean (and the same goes for most petroglyphs). There are also interesting petroglyphs in Brownstone Canyon (Figure 93).

Brownstone Canyon is accessible through a strenuous five mile round trip hike north through upper Gateway Canyon from the Sandstone Quarry.

Figure 97- Pictograph panel in Brownstone Canyon. Notice "wine glass" pictographs. This is probably the best pictograph panel in Southern Nevada. Panel is about six feet wide.

Figure 98 - Pictograph panel 2 in Brownstone Canyon. Panel is about ten feet wide. The pictographs in Brownstone Canyon are a priceless treasure that must be protected from possible vandalism by encroaching home development.

Willow Springs

Pictographs are found on both the north and south canyon walls of the Willow Springs picnic area. This was a major Native American encampment for thousands of years as springs emerge along the Aztec Sandstone/Kayenta Formation contact. Rainfall recharge seeps down through the joints (cracks) and through the pore spaces in the sandstone until it hits the low permeability red Kayenta shale, and emerges as springs. The pictographs on the south side (Figure 99) are accessible through the Petroglyph Wall Trail that begins near the end of the paved road and crosses the wash. (The nearby petroglyphs are shown in Figure 92.)

The handprint pictographs are seen by thousands of visitors each year (Figure 100). They are located on the north side of the road that enters the Willow Springs area from the Scenic Loop, to the east of the picnic area. The pictographs are located about six feet above ground surface on a large block of sandstone that fell from the cliff above. Just to the east of the hand prints is an agave roasting pit (Figure 101). There is also a pictograph in the rock shelter across the canyon (Figure 102). Pictographs are found in a few other places in Red Rock Canyon besides Willow Springs and Brownstone Canyon, including Lone Grapevine Spring.

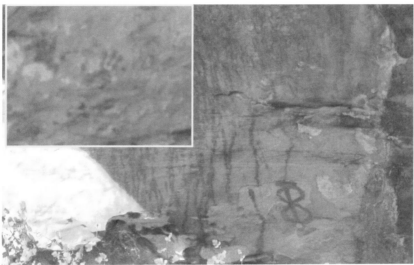

Figure 99 - Pictographs on south bedrock wall at Willow Springs near the end of the Petroglyph Wall Trail. The pictograph that resembles a vertical infinity symbol is about one foot long. In the inset are hand prints located nearby.

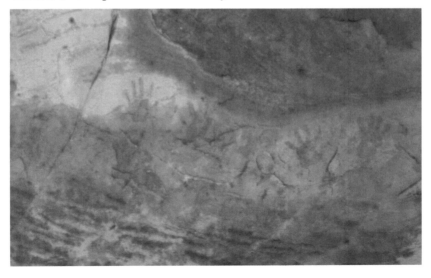

Figure 100 - Handprints on block of sandstone that fell from the cliff at the east end of the Willow Springs picnic area. These prints are seen by thousands of visitors to Red Rock Canyon every year.

Agave Roasting Pits

Native Americans who lived in Red Rock Canyon built agave roasting pits (Figure 101). In these pits, along with agave stalks, they roasted any animals they could kill for food, including bighorn sheep, rabbits, lizards, and tortoises. Eating agave stalks is supposed to be somewhat like eating artichokes. (Tequila is made from a Mexican agave.)

They made kind of a Dutch Oven by placing pieces of limestone below and on top of the food along with fuel (wood and dried plants). When the limestone became heated it was able to cook the food for hours. The limestone is altered by the heat and loses its ability to retain heat after being used one or two times. After the feast the discarded limestone and ash was thrown out of the pit, slowly building its distinctive donut or volcano appearance after many generations of use. Unburnt limestone is gray, while burnt limestone is white. White burnt limestone along with black ash is found on the slopes of the roasting pits. Figure 101 shows an agave roasting pit east of the Willow Springs picnic area, adjacent to the hand print pictographs (Figure 100).

This practice continued for many hundreds and perhaps thousands of years. The agave roasting

Figure 101 - Agave roasting pit east of Willow Springs picnic area. Agave roasting pits can be recognized by the white burnt limestone and black ash in the slopes.

Figure 102 - Rock shelter on Lost Creek - Children's Discovery Trail. Note pictograph below overhanging rock.

pits are believed to have been used for ceremonial feasts in addition to family tribe dinners. Some of the agave roasting pits are more than ten feet in diameter with the ring of spent white limestone and black ash over five feet high in places. About 145 agave roasting pits have been found throughout Red Rock Canyon.

Rock Shelters

Rock shelters are bedrock ledges beneath which the Indians took shelter from the elements. Unlike a cave, which is a large deep hole in the rock, a rock shelter is not completely enclosed by rock. A rock shelter with a pictograph can be seen on the Lost Creek - Children's Discovery Trail (Figure 102).

Please Don't Forget

For both Red Rock Canyon National Conservation Area (NCA) and Spring Mountain Ranch State Park, removing, disturbing or damaging any historic structure, artifact, rock, plant, animal, fossil or other feature is prohibited. Federal and State laws protect this area and its resources (Figure 103). And in Spring Mountain Ranch you can be issued a citation by the ranger if you go off trail.

If you find an unusual fossil, including tracks in the rock, or an artifact, photograph it, GPS the location if you can, and report the find to the rangers in the parks. The phone number for the rangers at the NCA is (702) 515-5350, and for the State Park is (702) 875-4141.

Figure 103 - Sign warning visitors not to damage or take artifacts. The signs are found in many places in the National Conservation Area. This one is at Lone Grapevine Spring.

REFERENCES

Axen, G.J., 1985, Geologic Map and Description of Structure and Stratigraphy, La Madre Mountain, Spring Mountains, Nevada: Geological Society of America Map and Chart Series MC-51.

Blakey, R.C., and Ranney, W., 2008, Ancient Landscapes of the Colorado Plateau. Grand Canyon, AZ: Grand Canyon Association.

Brean, Henry, 2011, Dinosaurs Hiked Red Rock Canyon: Las Vegas Review Journal, Nov. 21, 2011.

Burchfiel, B.C., Fleck, R., Secor, D.T., Vincelette, R. R., and Davis, G.A., 1974, Geology of the Spring Mountains, Nevada: Geological Society of America Bulletin, v. 85, p. 1013-1022.

Burchfiel, B.C., Cameron, C.S., and Royden, L.H., 1997, Geology of the Wilson Cliffs-Potosi Mountain Area, Southern Nevada: International Geology Review, v. 39, Issue 9.

Chan, M.A., and Parry, W.T., 2002, Rainbow of Rocks: Utah Geol. Survey Public Info. Series 77.

DeCourten, Frank, and Biggar, Norma, 2017, Roadside Geology of Nevada: Mountain Press, Missoula, Montana.

Dutch, Steven, 2005, Geologic Map of Red Rock Canyon, Nevada: https: // www. uwgbedu /dutchs/ VTrips/RedRx2005.HTM

Last, F.M., Last, W.M., and Halden, N.M., 2012, Modern and Late Holocene Dolomite Formation: Manito Lake, Saskatchewan, Canada: Sedimentary Geology, Vol. 281.

Longwell, C.R., Pampeyan, E.H., Bowyer, Ben, and Roberts, R.J., 1965, Geology and Mineral Deposits of Clark County, Nevada: Nevada Bureau of Mines and Geology Bulletin 62.

Marzolf, J.E., 1988, Reconstruction of Late Triassic and Early and Middle Jurassic Sedimentary Basins: Southwestern Colorado Plateau to Eastern Mojave Desert: *in* This Extended Land - Geological Journeys in the Southern Basin and Range, Field Trip Guidebook, Geological Society of America, Cordilleran Section Meeting, Las Vegas, NV.

Moulin, Tom, 2013, Red Rock Canyon Visitor Guide:Snell Press, Las Vegas, Nevada.

Nielson, R.L., 1983, Origin of Chert in Permian System in Southwestern Utah and Northwestern Arizona, American Association of Petroleum Geologists, v. 67, Issue 3.

Page, W.R., Dixon, G.L., and Workman, J.B., 1998, The Blue Diamond Landslide - A Tertiary Landslide

Breccia Deposit in the Blue Diamond Area, Southern Nevada *in* Seismic Hazards in the Blue Diamond Area, Southern Nevada. Nevada Bureau of Mines and Geology, Open File Report 98-6.

Page, W.R., Lundstrom, S.C., Harris, A.G., Langenheim, V.E., Workman, J.B., Mahan, S.A., Paces, J.B., Dixon, G.L., Rowley, P.D., Burchfiel, B.C., Bell, J.W., and Smith, E.I., 2005: Geologic and Geophysical Maps of the Las Vegas 30' x 60' Quadrangle, Clark and Nye Counties, Nevada, and Inyo County, California. U.S. Geological Survey Scientific Investigations Map 2814.

Peck, J.H., 1997, Detached and Rotated Landslide in Red Rock Canyon, Clark County, Nevada (Abs.): Association of Engineering Geologists, 40th Annual Meeting, Portland, Oregon.

Peck, J.H., 2000, Turtlehead Landslide - a Tertiary Landslide in Southern Nevada: *in* Geology of the Las Vegas Area, Clark County, Nevada; Bill Goodman, Ed., South Coast Geological Society Field Trip Guidebook, No. 28.

Rowland, Stephen, Breithaupt, B.H, Stoller, H.M., N.A. Matthews, and Saines, Marvin, 2014, First report of Dinosaur Synapsid, and Arthropod Tracks in the Aztec Sandstone (Lower-Mid Jurassic) of Red Rock Canyon National Conservation area, Southern Nevada: *in* Lockley, M.G. and Lucas, S.G. *eds.* Fossil Footprints of Western North America, NMMNHS Bull. 62.

Sahney, S., and Benton, M.J., 2008, Recovery From the Most Profound Mass Extinction of All Time: Proceedings of the Royal Society B.272 (16736): 759-765.

Saines, Marvin, and King, J.L., 2013, A Balanced Look at Precariously Balanced Rocks, With Examples from Red Rock Canyon National Conservation Area, Nevada (abs): Association of Environmental and Engineering Geologists 47th Annual Meeting, Seattle, Washington.

Tingley, J.V., Purkey, B.W., Duebendorfer, E.M., Smith, E.I., Price, J.G., and Castor, S.B., 2001, Geologic Tours in the Las Vegas Area: Nevada Bureau of Mines and Geology, Special Publication 16.

ABOUT THE AUTHOR
Marvin (Nick) Saines

After receiving a Ph.D. in geology from the University of Massachusetts, Nick settled in Chicago where he was an international consultant, with projects in Guyana, Iran, Afghanistan, Jordan, Malaysia, and the Sultanate of Oman. He moved to Las Vegas in 1989 to work on the Yucca Mountain Project, and then worked for local engineering/environmental consulting companies, with FEMA, and with the U.S. Army Corps of Engineers in Afghanistan. From 2011 to 2014 he was an interpretive naturalist/geologist with the Red Rock Canyon Interpretative Association (now Southern Nevada Conservancy). Many of the discoveries described in this book were made by the author and his colleagues during that period.

Nick is an adjunct professor of Environmental Science at the Community College of Southern Nevada, founder and Vice Chair of the Southern Nevada Chapter of the Association of Environmental and Engineering Geologists (AEG), and Southwest Regional Director for the AEG. He leads Sierra Club hikes in Red Rock Canyon and throughout the region, does hydrogeological consulting, and does geological research on the Jurassic sandstone.

He wrote and published a children's book, inspired by his discovery of a dinosaur footprint in the Jurassic sandstone, entitled "A Dinosaur Lives in Red Rock Canyon," which is available at the gift shop in Red Rock Canyon National Conservation Area, and at the Las Vegas Natural History Museum.

INDEX

aeolian deposits, 18
agave roasting pits, 22, 132, 142–144
alluvial deposits, 6, 52
alluvial fans, 72, 108
alluvium, 15–17, 45, 69, 124
ancient landslides, 101
Appalachian Mountains, 72
arachnids, 19
archaeology, 132–145
Arizona, 14, 54
Arnight Trail, 64, 111
Ash Creek Canyon, 117–118
Association of Environmental and Engineering Geologists (AEG), 115, 150
Aztec Sandstone, 5–6, 18–20, 26, 31, 44, 58–60, 62, 64, 66, 69–94, 98, 100, 106, 113, 117, 120–121, 127, 140

Bahamas, 25
Bering Land Bridge, 132
Bermuda, 25
bipedal dinosaur, 19, 87
Blue Diamond, 4, 6, 8, 13–14, 21, 28, 34–37, 40–41, 73, 101–103, 135
Blue Diamond Hill, 6, 13–14, 28, 34–37, 40–41, 73, 102
Blue Diamond Landslide, 21, 101–103
body fossils, 87

Bonanza King Formation, limestone, 23–24, 121
Bonnie Springs development (The Reserve at Red Rock Canyon), 1
boulder levees, 108–109
brachiopods, 28, 30
Brasilichnium, 90–91
breccia, 6, 23, 34–35, 38, 98, 101–105, 127–128, 131
Brownstone Canyon, 1, 19, 98, 100, 104–105, 115–116, 133–140
Brownstone Canyon pictographs, 134–139
bryozoans, 28
Bureau of Land Management (BLM), 4, 9–10, 19, 80–81, 85, 89–91

calcium carbonate, 18
Calico Basin, 6, 16, 20–21, 35, 45, 58, 60–63, 78–80, 84, 101, 104–105, 117–118, 120, 127
Calico Basin springs, 117–118, 127
Calico Hills, 13, 20, 79–80, 91–93, 115–120, 125, 127, 133
Calico Hills (Leifheit) dinosaur track Site, 91–93
Calico Hills Thrust Fault, 127
Calico Hills Trail, 115–116
California, 55, 62, 64, 70, 72,

122
Canada, 20, 66, 123
carbon dioxide, 32
carbonate cap (Kayenta Fm.), 1, 10, 16, 59, 61, 66–67
Cedar City, 106
chalcedony, 62–64
chert, 6, 29, 31
Children's Discovery Trail, 143–144
Chinle Formation, 1, 6, 16, 21, 44, 46, 50–58, 71, 98, 106
Coelophysis, 88
colluvium, 23
Colorado Plateau, 14
Conglomerate of Brownstone Basin, 6, 19, 98–100, 128
Coral Pink Sand Dunes State Park, 72
corals, 28
Cottonwood Fault, 20, 121, 130–131
Cottonwood Springs, 135
cowboy petroglyphs, 135–137
Cowboy Trail Rides, 28
Cretaceous extinction, 32
crinoids, 28, 30
cross bedding, 39–40, 73–74

Dale's Trail, 113
debris flow, 4, 6, 15, 17, 21, 101, 108–110
desert dinosaurs, 87–88
desert marbles, 78
desert varnish, 21–22, 29, 76, 80, 83, 85, 91–94, 119–120, 132–133
diagenesis, 29
Dinosaur Discovery Site, 58
dinosaur footprints/tracks, 1, 4, 19, 80–95, 119
dinosaurs, 3, 14, 19, 32, 52, 72, 82–88, 95, 97
Dixie State University, 10, 66

early mammals, 19, 88
East Point, 115–116, 120, 133
Echo Canyon, 28
erg, 72

Falling Man petroglyph, 135–137
Farallon Plate, 20
fault breccia, 23, 98, 127–128, 131
fenster, 19–20, 76–77, 98, 121, 127–131
First Creek, 42–43, 48–49, 108
Five Stop Hill, 127
fluvial deposits, 73–74
Fossil Ridge Trail, 38
fossils, 1, 12–16, 28–30, 33, 41, 44, 58, 61, 66, 82, 87, 99, 101, 104, 154
freshwater limestone, 48–49

Gateway Canyon, 78–79, 138
geologic cross section, Spring Mtn. Ranch, 4, 7

Geologic Investigative Team, 9, 49, 56, 62
geologic map, 4–5, 101–102, 121
Geologic Time Scale, 3
Geological Society of America field trip guidebook, 68
Glacial Maximum, 118
goethite, 76, 129, 138
Gold Butte National Monument, 135, 137
Grallator, 9, 19, 86–93
Grand Canyon, 14
granite batholith, 20, 122
Great Dying, The, 14–15, 32, 41
Great Jurassic Sand Sea, 18, 70, 72
Great Unconformity, 9
greenhouse gasses, 32
Grotto, The, 48–49
groundwater, 24–25, 29, 50, 65, 69, 75–79, 129
gypsum, 6, 14–15, 102

hematite, 18, 75–76, 128, 138
Hermit Formation, 13–14
Holocene, 106, 108, 113
Hopi indians, 78

Ice Age, 117, 132
Ice Box canyon, 16
Indian Marbles, 19, 78
interdune carbonates, 66
iron concretions, 19, 78–79

iron fixing bacteria, 21
iron oxide, 18–19, 76, 78, 83, 119, 128–129, 132, 138

jasper, 45–46
Johnson Farm, 58
Juniper Canyon, 1, 21, 108–112

Kaibab Formation, limestone, 15, 28–31, 33, 37–38, 41
Kayenta Formation, sandstones, shales, 16, 58, 71, 106
Kayenta-Moenave Formation, 6, 16, 58
Keystone Thrust Fault, 4, 6, 12, 20, 23, 26, 28, 45, 47, 70–71, 77, 121–124, 127–129
Keystone Thrust Fault trail, 77, 129
Knolls Trail, 111
Kraft Mountain, 62, 64

La Madre Fault, 20, 31, 125–126, 130
La Madre Mountain, 12–13, 35, 125
landslides, 4, 21, 101–106, 113
Las Vegas, 1, 9, 19, 51, 57, 66, 72, 85–89, 117, 132, 135
Las Vegas Natural History Museum, 57

Las Vegas Review Journal, 85–86
Las Vegas Springs, 135
Las Vegas Valley, 9, 72, 132
limestone conglomerate, 117
limonite, 76, 129, 138
lithification, 15, 18, 29, 72
lithified sand dunes, 18, 75
Little Red Rock, 2
Location map, 2
Lone Grapevine Spring, 135, 137, 140, 145
Lost Creek canyon, 16
Lower Red Member of Moenkopi Formation, 15, 17

mammal tracks, 90–92
manganese, 21, 83, 119, 132
meat-eating dinosaur, 19, 72
Megapnosaurus, 19, 87–88
Mescal Range, 72
methane, 32
methane hydrate deposits, 39
Metoposaur, 16, 56–57
Mexico, 45, 133, 136
Moenave equivalent, 16, 58
Moenave Formation, 6, 16, 58, 82
Moenkopi Formation, 6, 15–17, 33–34, 42–43, 48, 69, 71, 75–77
Moenkopi Ridge, 41–42
Moenkopi Trail, 41
Mojave Desert, 26
Moqui Marbles, 78, 19, 78
Mormon Road, 135, 137

Mormon Trail, 135
Mountain Springs, 48–49, 130, 135
Mt. Wilson, 47, 106
mudcracks, 42–43
Muffins, The, 35–38
Mushroom Rock, 115–116

Native-American archaeological features, 4, 22, 119, 132–145
natural dam in Juniper Canyon, 1, 108, 110, 112
Navajo Sandstone, 72, 82
New Mexico, 45
North American Plate, 20, 122
Nugget Sandstone, 72

Oak Creek, Oak Creek Canyon, 6, 15–17, 42–45, 50, 55, 77, 108–111
Octopodichnus, 84, 87, 89, 92–93
Old Spanish Trail, 130, 133, 135–137
opal, 62, 64
oxidized iron, 75–77

Pahrump, 130
paleogeographic map, 37, 39–40, 99–100
Paleozoic carbonates, 26, 28, 122
Permian Redbeds, 6, 13–14
Permo-Triassic extinction, 32

Petrified Forest Member, 6, 16, 21, 44, 54–57, 71, 106
Petrified Forest National Park, 54
petrified wood, 6, 16, 44, 50–51, 54
Petroglyph Wall Trail, 133–134, 140–141
petroglyphs, 4, 22, 119–120, 132–138, 140
pictographs, 4, 22, 132, 134, 138–142
Pine Creek dinosaur track sites, 89–93
Pine Creek, Pine Creek Canyon, 62, 64, 73–74, 80, 89–93, 108–113
Pleistocene, 106, 117–118
Potato Knoll, 21, 45–48, 50–55, 106–109, 113
Precambrian, 6, 19
Precariously balanced rocks (PBRs), 1, 21, 115–116
proto-mammal tracks, 19, 92

quartz, 45, 50, 62–64
quartzite, 6, 98
Queantoweap Sandstone, 13–14

Rainbow Mountain Wilderness, 23, 89
Red Rock Canyon Interpretive Association. *See Southern Nevada Conservancy*
Red Rock Canyon Visitor Center, 28, 31, 41–42, 80, 96, 121, 125
Red Rock Canyon Visitor Guide, 26
Red Rock Overlook, 41, 135
Red Spring, 117–118, 133–134
Red Spring Thrust Fault, 20, 26, 121–124, 127–128
reversal of topography, 104, 125–126
ripple marks, 42–43
Rock Canyon Conglomerate, 15, 34
rock shelters, 22, 132, 144
Rocky Gap Road, 12, 23
Royal Tyrrell (Drumheller) Museum, 52

Sahara Desert, 72
Saines dinosaur track site, 85–86
sand dunes, 18, 69–76, 82, 84, 87–88
Sandstone Quarry, 12, 115, 138
Santa Fe, 133
Scenic Loop/Drive, 1, 12–13, 20, 31, 42, 74, 98, 111, 125, 128, 140
Search and Rescue, 96
seismic activity, 1, 101, 115
Sevier Orogeny, 20, 122
shadow biosphere, 120
Shinarump Cliffs, 44
Shinarump Conglomerate,

155

16, 44–56, 59, 71, 106
Siberian eruptions, 32
Sierra Nevada, 20, 122
silica, 18, 29, 50, 62, 64
silica-rich groundwater, 29
siliceous layer, 1, 16, 62–65
Skull Rock landslide, 113–114
slump block, 21, 55, 106
Smith dinosaur track site, 81–84
SMYC Trail, 16, 62, 64, 66–67
Southern Illinois University, 68
Southern Nevada Conservancy, 9, 81, 92
sponges, 28, 30
spring deposits, 117–118
Spring Mountain Ranch State Park, 1, 4, 6–7, 12, 16, 44–46, 54–62, 68–71, 121, 144
stratigraphic table, 4, 6, 26
stromatolite, 24–25
sturzstroms, 101
subduction, 20, 122–123
sulfur dioxide, 32
Sultan Limestone, 6, 26–27

talus, 6, 16, 58
Tertiary landslides, 101
The Reserve at Red Rock Canyon, 1
therapods, 72, 87
Timpoweap Conglomerate, 9, 15, 34–40, 73
Toroweap Formation, 6, 14

Toroweap-Kaibab Formation, 14
Triassic channels, 1, 28, 37–40
tridactyl dinosaurs, 84
tufa, 117–118
Tule Springs Fossil Beds National Monument, 117
Turtlehead Landslide, 21, 104–105
Turtlehead Mountain Fault, 27
Turtlehead Peak, 11–12, 26–28, 135–136

University of Nevada, Las Vegas (UNLV), 9–10, 37, 55–56, 64, 66, 68, 87
University of Utah, 10, 75
Upper Gateway Canyon, 138
Upper Red Member of the Moenkopi Formation, 15–17, 42–43, 69, 71, 75–77
Utah, 10, 15–16, 34, 44, 55, 58, 61, 66, 72, 75, 82, 106, 123

Valley of Fire State Park, 19, 66, 72, 99, 123
vertebrate fossils, 1, 44
volcanic ash, 6, 55, 64
volcanic deposits, 68
volcanoes, 55, 62, 64, 122

wagon petroglyph, 137

Waterfall Canyon, 4, 12, 23–25
White Rock Hills, 16, 45, 125
White Rock Loop, 45, 47, 124
Willow Springs, 23, 80, 94, 133–143
Willow Springs dinosaur track site, 94, 134, 141
Willow Tank Formation, 19, 99
Wilson Cliffs, 15, 18, 41, 45, 69–71, 101, 106, 110, 121, 123, 131
wine glass pictographs, 139

X-ray diffraction, 10

Zion National Park, 72, 106

Made in the USA
Columbia, SC
11 July 2024